THE UNOFFICIAL GUIDE

TO COSMOS

THE UNOFFICIAL GUIDE TO COSMOS

FACT AND FICTION IN
NEIL deGRASSE TYSON'S
LANDMARK SCIENCE SERIES

DAVID KLINGHOFFER, EDITOR

DISCOVERY INSTITUTE PRESS   SEATTLE   2014

Description

In this unofficial viewers' guide to astrophysicist Neil deGrasse Tyson's reboot of the classic *Cosmos* TV series, contributors dissect each episode and explain where Tyson turns from objective science to science-flavored and fact-challenged preaching. Students, parents, and teachers will find this lively compendium a useful counterpoint to the new *Cosmos* series.

Copyright Notice

Publisher's Note

This book is part of a series published by the Center for Science & Culture at Discovery Institute in Seattle. Previous books include *Science and Human Origins* by Ann Gauger, Douglas Axe, and Casey Luskin; *The Myth of Junk DNA* by Jonathan Wells; *The Deniable Darwin & Other Essays* by David Berlinski; and *The Magician's Twin: C.S. Lewis on Science, Scientism, and Society*, edited by John G. West.

Library Cataloging Data

The Unofficial Guide to Cosmos: Fact and Fiction in Neil deGrasse Tyson's Landmark Science Series
Edited by David Klinghoffer. Contributions by David Klinghoffer, Jay Richards, Casey Luskin, and Douglas Ell.
182 pages, 6 x 9 x 0.4 inches & 0.6 lb, 229 x 152 x 1.1 cm. & 0.3 kg
Library of Congress Control Number: 2014952273
BISAC Subject: SCI015000 SCIENCE / Cosmology
BISAC Subject: SCI034000 SCIENCE / History
BISAC Subject: SCI075000 SCIENCE / Philosophy & Social Aspects
ISBN-13: 978-1-936599-22-6 (paperback)

Publisher Information

Discovery Institute Press, 208 Columbia Street, Seattle, WA 98104
Internet: http://www.discoveryinstitutepress.com/
Published in the United States of America on acid-free paper.
First Edition, First Printing. October 2014.
Cover Design: Brian Gage
Interior Layout: Michael W. Perry

Contents

INTRODUCING COSMOS

1.

WHY THIS SUBVERSIVE BOOK?

David Klinghoffer

COSMOS: *A SPACETIME ODYSSEY* IS THE 2014 REIMAGINING OF CARL Sagan's influential and esteemed public-television documentary series, *Cosmos: A Personal Voyage* (1980). There seems every reason to think that with the highly regarded astrophysicist Dr. Tyson as its genial host, graced by accessible science, gorgeous and expensive animation, critical acclaim, 12 Emmy Award nominations, and all around top-notch entertainment value, the new *Cosmos* will prove to be a fixture in public and private schools, used as a welcome supplement by science teachers seeking to enliven student interest in cosmology, biology, and more.

We present *The Unofficial Guide to Cosmos* as an aid to students, parents, and teachers seeking another perspective on the profound questions raised in the series by Dr. Tyson, questions that *Cosmos* seeks to answer in an often alarmingly one-sided fashion. The program pursues a definite philosophical agenda without ever once acknowledging that fact. It should not go unchallenged, even at the risk of appearing disrespectful of authorities like Dr. Tyson and Dr. Sagan.

Both the original and the revival are 13 episodes in total. During those weeks that the new series first aired, a question that was the subject of water-cooler discussion in our offices was: What is *Cosmos* about?

As the title suggests, it offers what is very broadly speaking a tour of the cosmos and a retelling of the story of how science came to understand man's place in the universe. But an odd thing about the program—very expensively produced and broadcast by the Fox Network and the National

Geographic Channel—is that it follows no obvious organizational scheme. In all our coverage of the series, we could identify no rhyme or reason to why one subject followed another, either at the level of an individual episode or of the entire 13-hour series.

In fact, only one episode was clearly focused on a particular topic, Episode 12, about global warming, which itself *seems* at first glance to have no connection to the subject of the series.

But it does. *Cosmos* makes sense once you understand that it comprises an instructional course in a way of conceiving the world, a prestigious one in academia but highly corrosive in our culture. This worldview has no completely satisfactory name, but many would call it "materialism," premised on the idea that material stuff is ultimately all there is in existence. In the famous words of astronomer Carl Sagan in the original series, "The cosmos is all that is, or ever was, or ever will be."

Under a materialist view, spirit, soul, ideas, information have no real existence. The universe and all the life in it arose by blind, unguided material processes. Religious faith is always a hindrance and never an inspiration to scientific investigation. The opposite view, going back millennia to ancient Greece, the Bible, through to the rise of modern science and beyond, is associated with the proposition that the cosmos gives evidence of intelligent design (ID). This design-based perspective argues that matter derives its existence from an underlying immaterial reality: wisdom, providence, design, information, call it what you will. Intelligent-design advocates including Stephen C. Meyer and William Dembski have argued that both at the origin of the universe and through the history of life, information not matter has come first.

On initially encountering them, these sound like abstract ideas. However, their real-world impact couldn't be more concrete or intimate. Materialism undergirds a widespread prejudice that equates humans with animals. It denies there is anything exceptional about the human race or planet Earth, that we have any appointed destiny, mission or purpose. Many of its adherents find the advancement of human settlement of the globe to be itself an offense against nature and our equals in the animal world. Hence the preoccupation with nightmare climate-change scenarios, casting human beings as a proliferating plague on the Earth.

Exploring the relative scientific and intellectual merit of these competing views of the universe—intelligent design versus scientific materialism—and illuminating their cultural consequences is the primary work of Discovery Institute's Center for Science & Culture.[1] It was only natural that our scientists and scholars followed the rollout of *Cosmos* with interest and concern.

There is much to debate about the content of the rebooted *Cosmos*, not least considering that even supporters of the program in the academic world acknowledged that its writers and producers sought to advance their argument by playing loose with the facts—indulging in what one friendly historian of science called "taradiddles," an innocent-sounding word that in fact means a deliberate untruth, a lie (see Chapter 28).

Is the word "lie" too strong in this context? We try to extend charity to all and so avoid, if possible, accusing others of "lying." It seems more likely that Dr. Tyson is misled by his own preferred narrative, the story he tells himself about how the world works. The story determines what does and doesn't count as a fact.

Indeed it seems relevant to point out that as we were putting this book together another critic, Sean Davis at *The Federalist*, noted various blunders and fabrications in Tyson's public presentations on political and other matters unrelated to the subject matter of *Cosmos*. Those included one egregious whopper, repeatedly employed in speeches, in which Tyson falsely depicted President George W. Bush as making a particularly foolish and biased comment immediately following the attacks of September 11, 2001.[2]

That aside, in this book you will find a collection of our responses to the *Cosmos* series, gathered mainly from Discovery Institute's online voice, *Evolution News & Views*, but including essays as well from *The Federal-*

1. Discovery Institute, Center for Science & Culture, http://www.discovery.org/id/about/.
2. Sean Davis, "Another Day, Another Quote Fabricated by Neil deGrasse Tyson," *The Federalist* (16 September 2014), http://thefederalist.com/2014/09/16/another-day-another-quote-fabricated-by-neil-degrasse-tyson/.

ist (Chapter 25 by Jay W. Richards) and *The Blaze* (Chapter 29 by Casey Luskin). It is our hope that the information and perspectives here will inform useful discussion of this landmark series among educators, parents, students, and other thoughtful citizens, advancing the truth and exposing "taradiddles."

2.

Teachers Abuzz

Don't Doubt It: *Cosmos* Is Coming to a School Near You

David Klinghoffer

DESPITE ITS UNDISGUISED AXE-GRINDING, HISTORY-BEFOGGING, AND faith-baiting excesses, there's no question that the rebooted *Cosmos* series with Neil deGrasse Tyson will be turning up in classrooms as a supplement to science education. It is not simply a matter of my speculating. As the program drew to a close, the Internet was abuzz with talk by teachers and others who are excited about the prospect.

I can't blame them. If I were an eighth-grade science teacher I would be on the lookout for ways to excite students about my subject, and *Cosmos*, despite its egregious faults, is a crowd-pleaser.

Journalist Dan Arel at *Salon* is looking forward to it:

It is a safe bet to assume that the popular, critically acclaimed show will turn up in classrooms across the country, and why shouldn't it? Tyson does a great job of explaining science so that everyone can understand what makes science fun and exciting.

[...]

Teachers should be excited about the DVD set, and how it can aid in teaching such things as the cosmic timeline.[3]

---

3. Dan Arel, "Why Neil deGrasse Tyson has creationists so thoroughly petrified, *Salon* (22 May 2014), http://www.salon.com/2014/05/22/why_neil_degrasse_has_creationists_so_thoroughly_petrified_partner/.

Commenting on a Facebook post by Dr. Tyson, teacher Chelsea Atwell writes:

Neil deGrasse Tyson, *Cosmos* is amazing! **I can't wait to buy it on DVD so that I can show it to my 8th grade science class during our universe section!!** My students do a lab on the universal timeline (much like the calendar you speak of in *Cosmos*). I can't wait to add the explanation for *Cosmos* into the lab! Thank you for inspiring a new generation of scientists![4]

Heather Scoville, a high-school teacher in Iowa, blogged on *Cosmos*, including recaps and weekly worksheets. She enthuses:

Teachers looking for an excellent television show to help drive home various science information to your students should look no further. The television station Fox has done society a great service by putting *Cosmos: A Spacetime Odyssey* hosted by Neil deGrasse Tyson on the air. Tyson delivers the often complicated ideas in a way that all levels of learners can understand and still be entertained by the stories. **Episodes of this show make great supplements in the science classroom and also can be used as a reward or movie day.**[5]

A website offering free lesson plans for teachers, HotChalkLessonPlansPage.com, gives thoughts on "Five Ways Neil deGrasse Tyson and Bill Nye Can Help Teachers Improve Science Literacy."[6] Writer Monica Fuglei observes, "Students love a movie day and both Nye and Tyson have television shows teachers can share with students." Regarding *Cosmos*, she says "students will find it easy to explore scientific topics with the aid of Tyson's excellent storytelling and compelling visuals."

Jon Lisi of Hollywood.com warns that, on one hand, *Cosmos* "successfully merges education with entertainment." However, "we do have to wonder if there's danger in placing so many eggs in the basket of educational TV.

---

4. "Neil deGrasse Tyson," https://www.facebook.com/neildegrassetyson/posts/10152355246217952. Emphasis added.

5. Heather Scoville, "Cosmos Episode Viewing Sheet," *About Evolution*, http://evolution.about.com/od/Cosmos-Teaching-Tools/fl/Cosmos-Episode-8-Viewing-Worksheet.htm. Emphasis added.

6. Monica Fuglei, "Five Ways Neil deGrasse Tyson and Bill Nye Can Help Teachers Improve Science Literacy," *HotChalk Lesson Plans*, http://lessonplanspage.com/five-ways-neil-degrasse-tyson-and-bill-nye-can-help-teachers-improve-science-literacy/.

Can programs like *Cosmos* sufficiently substitute for classroom education, or do they merely complement more traditional pedagogical practices?"[7]

He concludes by recommending a balanced approach: Use *Cosmos* in your classroom but don't forget to have kids "read a textbook" as well:

> [W]hile *Cosmos* and similar entertaining educational programs shouldn't replace a traditional classroom education, teachers and school districts need to understand that **these programs can help their students gain a more complete education.** Instead of selecting one or the other as "better" or more "substantial," we should realize that both approaches are necessary and important. On Monday, a teacher should have students read a textbook and take notes, and on Tuesday, a teacher should screen an entertaining documentary or interesting Ted Talk that covers what the student read in a more captivating manner.

After watching the first episode of *Cosmos*, Michael d'Estries at Mother Nature Network predicted:

> While Carl Sagan premiered a similar "cosmic calendar" back in 1980, the one on display last night was really incredible. **I've no doubt this part of Cosmos will be replayed countless times in science classes around the world for decades to come.**[8]

Writer Lindsey Weber is confident:

> *Cosmos* will likely soon find its way into the middle school canon, and hopefully the young people of America will find deGrasse Tyson as charming as I do.[9]

Many of these folks wrote before it became clear just how ideology-driven *Cosmos* really is. Maybe they've changed their minds. In any case, let me emphasize that I'm not here to condemn any teacher who uses the series in her class. It's not easy getting kids or adults jazzed about science, and doing so is a laudable goal. But parents should probably take an active role in supplementing what their kids learn from a "supplement" like this.

---

7. Jon Lisi, "Can 'Cosmos' Be a Substitute for Classroom Education? *Hollywood.com* (9 April 2014), http://www.hollywood.com/news/tv/56856688/cosmos-classroom-education.

8. Michael D'Estries, "5 great moments from Episode 1 of 'Cosmos'," (10 March 2014), http://www.mnn.com/lifestyle/arts-culture/blogs/5-great-moments-from-episode-1-of-cosmos.

9. Lindsey Weber, "10 GIFs That Prove Neil deGrasse Tyson Rules the Cosmos, *Vulture* (15 April 2014), http://www.vulture.com/2014/04/neil-degrasse-tyson-cosmos-gifs.html.

They could do so with some of our chapters here pointing out the myths being floated in one episode of the series after another.

3.

Taking Aim

Cosmos Has Intelligent Design in Its Crosshairs

David Klinghoffer

In case you had any uncertainty about the new Cosmos series, executive producer Seth MacFarlane had Darwin skeptics and alternatives to Darwinian evolution very much in his crosshairs. This was a major and costly project, though Fox won't say how costly—so it's flattering in a way.[10] In an interview in the *Los Angeles Times*, MacFarlane said:

> We've had a resurgence of creationism and intelligent design quote-unquote theory. There's been a real vacuum when it comes to science education. The nice thing about this show is that I think that it does what the original *Cosmos* did and presents it in such a flashy, entertaining way that, as Carl Sagan put it in 1980, even people who have no interest in science will watch just because it's a spectacle. People who watched the original *Cosmos* will sit down and watch with their kids.[11]

More:

> You see it in the rise of schools questioning evolution, all these things piling up that betray the fact that we've lost our way in terms of our scientific literacy and it's incredibly damaging to the evolution of any society. I thought we solved this whole evolution thing years and years ago but I

---

10. David Itzkoff, "'Family Guy' Guy as Astrobiology Guy," *New York Times* (28 February 2014), http://www.nytimes.com/2014/03/02/arts/television/seth-macfarlane-champions-new-cosmos-series-on-fox.html.

11. Meredith Blake, "Seth MacFarlane hopes 'Cosmos' counteracts 'junk science,' creationism," *Los Angeles Times* (7 March 2014), http://www.latimes.com/entertainment/tv/showtracker/la-et-st-qa-seth-macfarlane-cosmos-science-creationism-20140305,0,4963186.story.

guess not, I guess it still needs to be explained. There are a number of areas where scientific illiteracy rears its head. I think in a lot of cases, it's not a conscious rejection, it's just that there's nothing out there that's feeding that hunger that maybe they don't know that they have.

I would bet that MacFarlane has no idea about the details of the challenge from intelligent design, whether in the realm of cosmology or biology, as most people who denounce ID in the name of "scientific literacy" do not. Today, materialism of the Carl Sagan variety is overwhelmingly maintained by a careful averting of the eyes from counterevidence and counterarguments. Steve Meyer's comment seems apt:

> The problem with materialists is they think that in [the brief span of the history of modern science], science has got all the mysteries of existence figured out.... In fact, we are just beginning to uncover the scientific evidence that the material cosmos is not all there is.[12]

12. "Discovery Institute Scholars Predict that, as a Vehicle for Materialism, the New *Cosmos* Will Be as Screechy as Sagan's Original," *Evolution News & Views* (6 March 2014), http://www.evolutionnews.org/2014/03/discovery_insti_5082921.html.

EPISODE 1: STANDING UP IN THE MILKY WAY

4.

PERSECUTING SCIENCE

COSMOS WITH NEIL DEGRASSE TYSON: SAME OLD PRODUCT, BRIGHT NEW PACKAGING

Casey Luskin

IF THERE WAS ANY QUESTION THAT THE COSMOS SERIES WOULD BE PO-litically charged and have a materialistic ideological message, consider what viewers saw in its first sixty seconds. The opening featured President Obama, with the Presidential Seal in the background, giving a statement endorsing the new series, praising "the spirit of discovery that Carl Sagan captured in the original *Cosmos*."

That's not necessarily bad. In fact it could have been a good thing, except for what happens next. *Immediately* following President Obama's endorsement, the show replays Carl Sagan's famous materialistic credo from the opening of the original *Cosmos* series, stating: "The cosmos is all that is, or ever was, or ever will be." Does it violate the separation of church and state for the President of the United States to be portrayed seemingly officially endorsing Sagan's philosophy? Is this what President Obama intended when he promised in his first Inaugural Address to "restore science to its rightful place"?[13] Or did the President simply give a general endorsement of the series, and the producers of *Cosmos* positioned it so as to appear that he endorsed Sagan's atheistic worldview?

---

13. Gardiner Harris and William J. Broad, "Scientist Welcome Obama's Words," *New York Times* (21 January 2009), http://www.nytimes.com/2009/01/22/us/politics/22science. html.

Whatever the answers, the show's intent to open with a heavy-hitting endorsement of materialism came through loud and clear. Ironically, viewers were immediately then told by the series's host, astrophysicist Neil deGrasse Tyson, that science follows a "set of rules." It should:

+ Follow the evidence wherever leads.

+ Question everything.

+ Put things to the test.

Again, that all sounds fine and good. But does science support Sagan's belief that "The cosmos is all that is, or ever was, or ever will be"? At best, that's a philosophical or metaphysical claim that goes *beyond* science. At worst, Sagan's claim is refuted by science, since known natural laws are incapable of explaining certain peculiar properties of the cosmos, including the life-friendly fine-tuning of the universe[14] and the fine-tuning of biological information to yield complex structures.[15] If the cosmos is "all there is," then the cosmos cannot account for its own existence, nor the complexity of what's inside it.

Before I launch into any more critiques, let me note some genuine positives about the series. First, the expensive CGI that animates the new *Cosmos* is easy on the eyes, and deliberately appeals to sci-fi fans like myself. Having watched every episode of every *Star Trek* series multiple times, I was excited to learn that the new *Cosmos* series was directed by Brannon Braga, who also helped create *Star Trek: The Next Generation*, *Star Trek: Voyager*, and *Star Trek: Enterprise*. In the first few minutes of *Cosmos*, Braga's influence is clear. Neil deGrasse Tyson is portrayed flying in a sleek spaceship through our solar system, the Milky Way galaxy, and then the entire universe, giving us a visually stunning and innovative tour of our "cosmic address," as Tyson puts it. That's another positive about the series: Tyson is a fabulous science communicator. If only he had used this series

---

14. "Evidence of the Design of the Universe through the Anthropic Principle," *IDEA*, http://www.ideacenter.org/contentmgr/showdetails.php/id/837.

15. Casey Luskin, "What Is a Proper Test of Intelligent Design?," *Evolution News & Views* (10 September 2011), http://www.evolutionnews.org/2011/09/what_is_a_proper_test_of_intel050631.html.

to simply communicate science, rather than science plus a heavy dose of materialist philosophy.

In the first episode, Tyson devotes lengthy segments to promoting the old tale that religion is at war with science and strongly promotes the idea that religion opposes intellectual advancement. He tells the story of the sixteenth-century cultist and philosopher Giordano Bruno, who he says lived in a time without "freedom of speech" or "separation of church and state," and thus fell into the clutches of the "thought police" of the Inquisition for disagreeing with the church's geocentric views. Of course the main religious authority of that time was the Catholic Church, and the program shows angry priests with evil-sounding British accents dressed in full religious garb throwing Bruno out on the street, and eventually burning him at the stake.

Just to make sure that other Christians who aren't Catholic also understand their religions too hinder scientific progress, Tyson goes out of his way to point out that Bruno was opposed by "Calvinists in Switzerland," and "Lutherans in Germany," including the great Protestant reformer Martin Luther himself. He never mentions that Protestants aren't the ones who burned Bruno at the stake, nor does he ever mention that most of the founders of modern science were Christians. But I digress...

It's a lengthy scene, all to highlight some of the darkest chapters of Christianity in Europe. But the entire retelling of Bruno's fate lasts a good portion of the first episode's hour. Why make religious persecution some four hundred years ago a major focus of a widely publicized television series that is ostensibly about promoting science?

Actually, I'd love to see a TV show aimed at helping the public to understand the dangers of hindering academic freedom for scientists. I suppose if you wanted to cover that topic, you'd want to talk about the evil things some members of the church did to persecute scientists hundreds of years ago. But why stop there? Why not also talk about how Lysenkoists[16] in the USSR persecuted scientists who didn't support their atheist, Com-

---

16. Casey Luskin, "Evolutionary Biologist Austin Hughes Praises Fine-tuning Arguments, Critiques Scientism," *Evolution News & Views* (14 December 2012), http://www.evolutionnews.org/2012/12/evolutionary_bi067491.html.

munist ideology during the twentieth century? Or why not talk about the numerous well-documented examples of scientists who have faced persecution and discrimination for disagreeing with Darwinian evolution in just the last few years? For example, in chronological order going back more than two decades, here's what has been happening.

1993

IN 1993, SAN FRANCISCO STATE UNIVERSITY BIOLOGY PROFESSOR DEAN Kenyon was forced to stop teaching introductory biology because he was informing students that scientists had doubts about materialist theories of the origin of life.[17]

1998

IN A SIMILAR CASE FIVE YEARS LATER, MINNESOTA HIGH SCHOOL TEACHer Rodney LeVake was removed from teaching biology after expressing skepticism about Darwin's theory.[18] LeVake, who holds a master's degree in biology, agreed to teach evolution as required in the district's curriculum, but said he wanted to "accompany that treatment of evolution with an honest look at the difficulties and inconsistencies of the theory."

1999

IN 1999, ID THEORIST WILLIAM DEMBSKI FOUNDED THE POLANYI CENter at Baylor University to allow scientists and scholars to conduct scientific research into intelligent design.[19] The Center was later shut down largely due to intolerance of ID among Baylor faculty.

---

17. Stephen C. Meyer, "Danger: Indoctrination," *Discovery Institute*, http://www.discovery. org/a/93. From *The Wall Street Journal* (6 December 1993).

18. Casey Luskin, "*LeVake v. Independent School District*: Administrators May Control the Evolution Curriculum," *Evolution News & Views* (29 January 2010), http://www. evolutionnews.org/2010/01/levake_v_independent_school_di031411.html.

19. William A. Dembski, "The Rise and Fall of Baylor University's Michael Polanyi Center," *Design Inference*, http://www.designinference.com/documents/2007.12.MPC_Rise_and_ Fall.htm.

2001

FROM 1998 TO 2002, WITH THE PEAK OF THE CONTROVERSY COMING IN 2001, Roger DeHart, a public high school biology teacher in Washington State, was denied the freedom to have his students read articles from mainstream science publications that made scientific criticisms of certain pieces of evidence typically offered to support Darwinian theory.[20] One of the forbidden articles was written by noted evolutionist Stephen Jay Gould. Although DeHart complied with this ban, he was later removed from teaching biology.

2003

IN 2003, MISSISSIPPI CHEMISTRY PROFESSOR NANCY BRYSON WAS ASKED by Mississippi University for Women to resign as head of the Division of Science and Mathematics after she gave a lecture to honors students called "Critical Thinking on Evolution." She remarked, "Students at my college got the message very clearly[:] do not ask any questions about Darwinism."[21]

2005

IN 2005, AN INVESTIGATION BY THE U.S. OFFICE OF SPECIAL COUNSEL (OSC), an independent federal agency, found that biologist Richard Sternberg[22] experienced retaliation by his co-workers and superiors at the Smithsonian, including transfer to a hostile supervisor, removal of his name placard from his door, deprivation of workspace, subjection to work requirements not imposed on others, restriction of specimen access, and loss of his keys. Why? Because he allowed a pro-ID article to be published in a biology journal. The OSC investigation concluded that the "Smithsonian's top officials permit[ed] the demotion and harassment of [a] scientist skeptical of Darwinian evolution" and "officials explicitly acknowledged in

---

20. Anika Smith, "Teaching the Controversy in South Korea: An Interview with Biology Teacher Roger DeHart," *ID the Future* (16 March 2011), http://www.idthefuture.com/2011/03/teaching_the_controversy_in_so.html.

21. Texas State Board of Education Hearing Transcript, 505 (September 10, 2003).

22. Richard Sternberg, "Smithsonian Controversy," *Richard Sternberg*, http://www.richardsternberg.com/smithsonian.php.

emails their intent to pressure Sternberg to resign because of his role in the publication of the [pro-intelligent design] Meyer paper and his views on evolution."

Also in 2005, Smithsonian spokesman Randall Kremer objected to a private screening of the pro-ID film *The Privileged Planet* because it drew a "philosophical conclusion."[23] The Smithsonian made no complaints when Sagan's original *Cosmos* in 1980 argued that "The cosmos is all that is, or ever was, or ever will be."

In 2005, over 120 faculty members at Iowa State University (ISU) signed a petition denouncing ID and calling on "all faculty members to… reject efforts to portray Intelligent Design as science." These efforts were significant not just because they opposed academic freedom by demanding conformity among faculty to reject ID, but because they focused on creating a hostile environment for pro-ID astronomer Guillermo Gonzalez, co-author of *The Privileged Planet*, who was denied tenure at ISU in 2006 due to his support for ID.[24] Both public and private statements exposed through public records requests revealed that members of ISU's department in physics and astronomy voted against Gonzalez's tenure due to his support for ID.

In 2005, the president of the University of Idaho instituted a campus-wide classroom speech-code, where "evolution" was "the only curriculum that is appropriate" for science classes.[25] This was done in retaliation against a professor at the university, Scott Minnich, who at the time was testifying in favor of intelligent design as an expert witness at the *Kitzmiller v. Dover* trial.

Also in 2005, Cornell's then-interim president Hunter Rawlings devoted a State of the University Address "to denounce 'intelligent design,'

23. Guillermo Gonzalez, "Misrepresenting intelligent design," *The Scientist* (29 August 2005), http://www.the-scientist.com/?articles.view/articleNo/16671/title/Misrepresenting-intelligent-design/.

24. Discovery staff, "Dr. Guillermo Gonzalez and Academic Persecution," *Discovery Institute* (8 February 2008), http://www.discovery.org/a/2939.

25. Robert Crowther, "AP Breaks Story on Academic Freedom Under Attack at U of Idaho," *Evolution News & Views* (7 October 2005), http://www.evolutionnews.org/2005/10/ap_breaks_story_on_academic_freedom_unde001069.html.

arguing that it has no place in science classrooms and calling on faculty members in a range of disciplines" to similarly attack ID.[26]

Again in 2005, top biology professors at Ohio State University derailed a doctoral student's thesis defense by writing a letter claiming "there are no valid scientific data challenging macroevolution" and therefore the student's teaching about problems with neo-Darwinism was "unethical" and "deliberate miseducation."[27]

The same year, pro-ID adjunct biology professor Caroline Crocker lost her job at George Mason University after teaching students about the evidence both for and against evolution and mentioning ID as a possible alternative to Darwinism.[28] While her former employer maintains that it simply chose not to renew her contract, she was specifically told she would be "disciplined" for teaching students about the scientific controversy over evolution.

2006

IN 2006, A PROFESSOR OF BIOCHEMISTRY AND LEADING BIOCHEMISTRY textbook author at the University of Toronto, Laurence A. Moran, stated that a major public research university "should never have admitted" students who support ID and should "just flunk the lot of them and make room for smart students."[29]

2007

IN 2007, ROBERT MARKS, DISTINGUISHED PROFESSOR OF ELECTRICAL and Computer Engineering at Baylor University, had established an Evolutionary Informatics Lab at Baylor to study the ability of Darwinian pro-

26. Scott Jaschik, "A Call to Action Against Intelligent Design," *Inside Higher Ed* (24 October 2005), http://www.insidehighered.com/news/2005/10/24/id.

27. John G. West, "Darwinists Continue Smear Campaign Against Ohio Grad Student," *Evolution News & Views* (18 July 2005), http://www.evolutionnews.org/2005/07/darwinists_continue_smear_campaign_again000979.html.

28. Casey Luskin, "Free to Think: Caroline Crocker's New Book Tells Story of Discrimination Against Intelligent Design," *Evolution News & Views* (15 July 2010), http://www.evolutionnews.org/2010/07/free_to_think_caroline_crocker036671.html.

29. Casey Luskin, "Leading Biochemistry Textbook Author: Pro-ID Undergraduates Should Never Have [been] Admitted,'" *Evolution News & Views* (20 November 2006), http://www.evolutionnews.org/2006/11/author_of_leading_biochemistry002876.html.

cesses to generate new information using computer simulations and evolutionary algorithms. However, after Dr. Marks was interviewed by *ID the Future* in 2007, he received a letter from his dean warning that the website was "associated" with "ID," and he was forced to take the lab's site down and move the lab itself off campus.[30]

True, ID-critics may not be burning people at the stake, but they have become so intolerant that in 2007, the Council of Europe, the leading European "human rights" organization, adopted a resolution calling ID a potential "threat to human rights"![31]

2009

IN 2009 THE STATE-FUNDED CALIFORNIA SCIENCE CENTER (CSC) MUSEum cancelled the contract of a pro-ID group, American Freedom Alliance (AFA), to show a pro-ID film. The lawsuit was settled in August 2011, with the CSC agreeing to pay AFA $110,000 to avoid a public trial. However, documents disclosed during the course of litigation showed that employees of the CSC, the Smithsonian Institution, and the Natural History Museum of Los Angeles County joined with other LA-area academics to suppress the expression of ID, most egregiously by pressing CSC decision-makers to hastily cancel AFA's event.[32]

In 2009, David Coppedge was demoted and punished for sharing pro-ID videos with co-workers at Jet Propulsion Lab. Later, his employment was terminated.[33]

30. Casey Luskin, "Banned Item of the Year: Dr. Robert Marks' Evolutionary Informatics Website," *Evolution News & Views* (3 October 2007), http://www.evolutionnews. org/2007/10/banned_item_of_the_year_dr_rob004303.html.

31. Casey Luskin, "European Darwinists Attempt to Criminalize Intelligent Design as a 'Threat to Human Rights,'" *Evolution News & Views* (1 August 2007), http://www. evolutionnews.org/2007/08/european_darwinists_attempt_to004047.html.

32. Casey Luskin, "Evidence Revealed in California Science Center Lawsuit Shows Intolerance and Efforts to Suppress Intelligent Design," *Evolution News & Views* (30 August 2011), http://www.evolutionnews.org/2011/08/evidence_revealed_in_californi050191.html.

33. Discovery staff, "Background on David Coppedge and the Lawsuit Against NASA's Jet Propulsion Laboratory," *Discovery Institute* (19 March 2012), http://www.discovery. org/a/14511.

2011

IN 2011, A BIOLOGY PROFESSOR AT THE UNIVERSITY OF WAIKATO STATED that "If, for example, a student were to use examples such as the bacterial flagellum to advance an ID view, then they should expect to be marked down."[34]

The same year Jerry Coyne, an evolutionary biologist at the University of Chicago, stated that "adherence to ID (which, after all, claims to be a nonreligious theory) should be absolute grounds for not hiring a science professor."[35]

In January 2011, the University of Kentucky (UK) paid over $100,000 to settle astronomer Martin Gaskell's lawsuit claiming that he was wrongfully denied employment because of his perceived doubts about Darwin. UK faculty admitted that Gaskell was the most qualified applicant for the position, but they hired a much less qualified candidate out of concerns about statements Gaskell had made that were critical of Darwinian evolution.[36]

In 2011, the journal *Applied Mathematics Letters* paid $10,000 and publicly apologized to avoid litigation after it wrongfully withdrew mathematician Granville Sewell's paper critiquing neo-Darwinism.[37]

2012

IN 2012, SPRINGER-VERLAG ILLEGALLY BREACHED A CONTRACT TO PUBlish the proceedings of an ID-friendly research conference at Cornell Uni-

---

34. Casey Luskin, "Want a Good Grade in Alison Campbell's College Biology Course? Don't Endorse Intelligent Design (Updated)," *Evolution News & Views* (2 March 2011), http://www.evolutionnews.org/2011/03/want_a_good_grade_in_allison_c044581.html.

35. Michael Egnor, "Jerry Coyne: '...Adherence to ID... Should Be Absolute Grounds for Not Hiring a Science Professor,'" *Evolution News & Views* (21 March 2011), http://www.evolutionnews.org/2011/03/jerry_coyne045001.html.

36. Casey Luskin, "E-mails in Gaskell Case Show That Darwin Skeptics Need Not Apply to the University of Kentucky," *Evolution News & Views* (10 February 2011), http://www.evolutionnews.org/2011/02/e-mails_in_gaskell_case_show_t043711.html.

37. John West, "Journal Apologizes and Pays $10,000 After Censoring Article," *Evolution News & Views* (7 June 2011), http://www.evolutionnews.org/2011/06/journal_apologizes_and_pays_10047121.html.

versity after a pressure campaign was mounted by pro-Darwin activists to have the book scuttled.[38]

2013

IN 2013, BALL STATE UNIVERSITY (BSU) PRESIDENT JO ANN GORA issued a speech code declaring that "intelligent design is not appropriate content for science courses" at BSU, after atheist activists from the Freedom from Religion Foundation charged that a "Boundaries of Science" course taught by a well-liked physics professor (Eric Hedin) violated the Constitution by favorably discussing intelligent design.[39]

Also in 2013, atheist activists forced Amarillo College to cancel an intelligent design course after they threatened disruption if it went forward.[40]

Freedom of Speech

SO IF NEIL DEGRASSE TYSON FELT SO STRONGLY THAT IT'S IMPORTANT to teach the public about the importance of "freedom of speech" and for scientists to "question everything," then why didn't he mention any of these recent incidents where skeptics of Darwinian evolution or proponents of intelligent design had their academic freedom violated? Why did he only focus on incidents from four hundred years ago, while ignoring all the numerous instances in the present day where atheist-Darwin activists have suppressed the rights of ID-friendly scientists? Could it be because Tyson himself is basically an atheist and sees *Cosmos* as a great opportunity to promote his materialistic worldview?

---

38. Casey Luskin, "On the Origin of the Controversy Over *Biological Information: New Perspectives*," *Evolution News & Views* (19 August 2013), http://www.evolutionnews. org/2013/08/on_the_origin_o_3075521.html.

39. John G. West, "Ball State President's Orwellian Attack on Academic Freedom," *Evolution News & Views* (1 August 2013), http://www.evolutionnews.org/2013/08/ ball_state_pres075041.html.

40. Casey Luskin, "Bullies-R-Us: How 'Freethought Oasis' Threatened 'Disruption' and Pressured a College into Canceling Intelligent Design Course," *Evolution News & Views* (10 December 2013), http://www.evolutionnews.org/2013/12/freethought-oasis- bullies079991.html.

Tyson may officially deny that he's an atheist, but that's just standard political posturing.[41] As he said at the "Beyond Belief" conference, which helped launch the New Atheist movement in 2006:

> I want to put on the table, not why 85% of the members of the National Academy of Sciences reject God, I want to know why 15% of the National Academy don't. That's really what we've got to address here. Otherwise the public is secondary to this.[42]

There's even a Facebook page created by fans of "Tysonism" which purports to promote "a secular religion based on the philosophy of astrophysicist Dr. Neil deGrasse Tyson." The page quotes him as saying things like:

> The more I learn about the universe, the less convinced I am that there's any sort of benevolent force that has anything to do with it, at all.[43]

Another sign that *Cosmos* has a materialistic agenda is the fact that its executive producer is celebrity atheist Seth MacFarlane (creator of *Family Guy*), who commented in an interview with *Esquire* about the need to be "vocal about the advancement of knowledge over faith":

> ESQ: … I see you've recently become rather vocal about your atheism. Isn't it antithetical to make public proclamations about secularism?
>
> SM: We have to. Because of all the mysticism and stuff that's gotten so popular.
>
> ESQ: But when you wave banners, how does it differ from religion?
>
> SM: It's like the civil-rights movement. There have to be people who are vocal about the advancement of knowledge over faith.[44]

Could the anti-religious message already seen in the first episode of *Cosmos* be MacFarlane's attempt to promote what he thinks is "the advancement of knowledge over faith"?

---

41. "Neil deGrasse Tyson: Atheist or Agnostic?," https://www.youtube.com/watch?v=CzSMC5rWvos.

42. Casey Luskin, "Rationalization in the Debate over Evolution," *Evolution News & Views* (11 December 2006), http://www.evolutionnews.org/2006/12/atheists_are_great_rationalize002941.html.

43. https://m.facebook.com/Tysonism.

44. Stacey G. Woods, "Hungover with Seth MacFarlane," *Esquire* (18 August 2009), http://www.esquire.com/features/the-screen/seth-macfarlane-interview-0909.

As David Klinghoffer showed in Chapter 3, *Cosmos* planned to take aim at intelligent design. Just how badly does *Cosmos* botch its attempts to attack intelligent design? Read on.

5.

BRUNO'S HERESIES

COSMOS REVIVES THE SCIENTIFIC MARTYR
MYTH OF GIORDANO BRUNO

Jay W. Richards

MANY VIEWERS MAY HAVE BEEN BAFFLED THAT SO MUCH TIME would be spent on Giordano Bruno, an Italian Dominican friar born in 1548 who was neither a scientist nor credited with any scientific discovery. Why is that? It's because he's the only one with even a passing association with a scientific controversy to be burned at the stake during this period of history. As a result, since the nineteenth century, when the mythological warfare between science and Christianity was invented, Bruno has been a leading character.

But there's one problem: Bruno's execution, troubling as it was, had virtually nothing to do with his scientific views. He was condemned and burned in 1600, but it was not because he speculated that the Earth rotated around the sun along with the other planets. He was condemned because he denied the doctrine of the Trinity, the Virgin Birth, and transubstantiation, claimed that all would be saved, and taught that there was an infinite swarm of eternal worlds of which ours was only one. The latter idea he got from the ancient (materialist) philosopher Lucretius. Is it any surprise, then, that, as a defrocked Dominican friar denying essential tenets of Catholic doctrine and drawing strength from the closest thing to an atheist in the Roman world, he might have gotten in trouble with the Inquisition? Yet a documentary series about science and our knowledge of the universe fritters away valuable airtime on this Dominican mystic and heretic, while scarcely mentioning Copernicus, the Polish guy who actually wrote the book proposing a sun-centered universe. Why?

The reason is obvious once you see that *Cosmos* is not just good ole science education, but rather a glossy multi-million-dollar piece of agit-prop for scientific materialism. The biography of Copernicus, whatever its scientific significance, provides precious little fodder of the desired kind. Copernicus died peacefully in his bed just as his book, *On the Revolution of the Heavenly Spheres*, was hitting the bookstores (such as there were in 1543). And his most famous disciple, Galileo, despite being censured by the Holy See, died peacefully as well. So it falls to Bruno, who had no scientific achievements, to stand in as a martyr for science. I'd venture that virtually no one other than scholars of Christian history would even know the name of Giordano Bruno but for the propaganda machine of scientific materialism, which needed a martyr for its meta-narrative.

Neil deGrasse Tyson does include a few hedges. While wandering the streets of modern-day Rome, he admits that Bruno wasn't a scientist and that his view of a sun-centered solar system was a "lucky guess." And during the animated dramatization of Bruno's sentence, the dark and menacing judge finds the brave Dominican guilty not just of being a Copernican, but of various theological trivialities which are never otherwise mentioned or explained. Despite these hints at nuance, not one viewer in a thousand could miss the real message: Christianity has been the enemy of science, and its henchmen tried to kill off the first brave souls who ventured a scientific thought.

What's most stunning is that the facts about Bruno are not exactly well-kept secrets. The Wikipedia entry for Giordano Bruno gets its more or less right (thanks to Steve Greydanus).[45] What that means is that the material-ist bias of the producers, editors, and writers of *Cosmos* is so complete that they couldn't be bothered even to check Wikipedia. One wouldn't want to let the facts get in the way of a good propaganda. The irony is that that makes it not-so-good propaganda.

---

45. "Giordano Bruno," Wikipedia, http://en.wikipedia.org/wiki/Giordano_Bruno.

6.

FLAWED HISTORY

COSMOS GETS SLAMMED FOR ITS INACCURATE AND REVISIONIST HISTORY OF GIORDANO BRUNO

Casey Luskin

IN CHAPTER 4, I MENTIONED THAT THE FIRST EPISODE OF COSMOS DEvotes an unusual amount of time—for a science show—promoting the old "warfare" model of science and religion, and the myth that religion has hindered the advancement of science. In Chapter 5, Jay Richards critiques Cosmos's revisionist history of Giordano Bruno, the cultist philosopher who was persecuted by the Catholic Church for his philosophy that (among other things) worshiped Egyptian deities. But Dr. Richards is not the only critic. In fact, a number of mainstream sources, *decidedly unfriendly to intelligent design*, have made similar criticisms.

The television website Zap2it.com criticizes Neil deGrasse Tyson for his "questionable history" of Bruno:

Unfortunately for Cosmos, Bruno wasn't terribly heroic. And he wasn't a scientist at all....

What Cosmos does not point out to its audiences is that the Catholic Church didn't really care about Bruno's views on the Earth moving around the Sun. His crimes—the ones for which he was executed—were theological. Several actual scientists in this period happily investigated the ideas of Copernicus' theories without running into trouble. Even Galileo only got in trouble when he published books that directly mocked the Church's adherence to the Earth being at the center.[46]

Why Does This Matter?

ZAP2IT.COM APTLY ASKS, "WHY BELIEVE THE SCIENCE IF OTHER PARTS of the show are inaccurate?" and comments about the anti-religious message of *Cosmos*:

> It's an unstated goal of *Cosmos* to champion science and scientific reasoning over superstition and religious dogmatism. But you're not going to win over anyone by vilifying religion in the face of science. Add in Bruno flying into space in an overtly crucifixion stance almost seems like giving religion the finger.[47]

At *Science 2.0*, Hank Campbell pointed out that *Cosmos* was wrong to claim Bruno got his ideas by reading "banned" books because Bruno's source, the Roman philosopher Lucretius, was widely read at the time, and people were "not martyred for reading it." He charges that *Cosmos's* treatment of Bruno was "flat-out incorrect" and "smacks of an agenda." Campbell explains that Bruno wasn't persecuted for his science, but for promoting cultic philosophical beliefs:

> Bruno's "science" was never mentioned during his trial, he was on trial for being a cult worshiper. He only took up the cause of Copernicus because he believed in the Egyptian god Thoth and Hermetism and their belief that the Earth revolved around the Sun, not because he had perceived anything radical. Galileo rightly dismissed most of Bruno's teachings as philosophical mumbo-jumbo. Bruno was only revived as a "scientist" and a martyr for science by anti-religious humanists in the 19th century....

> Bruno was not a martyr for science, the way the cartoon in *Cosmos: A Spacetime Odyssey* alleges, he was a martyr for magic. He actually was a heretic. Sorry, but 400 years ago when you repeatedly lecture about what

46. Laurel Brown, "'Cosmos' review: Neil deGrasse Tyson brings brilliant science, questionable history to the world," *Zap2it* (9 March 2014), http://blog.zap2it.com/frominsidethebox/2014/03/cosmos-review-neil-degrasse-tyson-brings-brilliant-science-questionable-history-to-the-world.html.

47. Ibid.

was regarded as a cult and insist Catholics and Protestants need to accept Hermetism[48] as fact, you are getting into trouble.[49]

Campbell concludes:

> It sets an unfortunate tone that they slipped revisionist history in with science—it is the story of Bruno as if it were written by a blogger on some "free thought" site. Are humanists and atheists the key market for this program?...
>
> [T]he Bruno story came across as more of a program Richard Dawkins would have hosted than Carl Sagan. And that's too bad, because Tyson is not divisive like that.[50]

Or, perhaps something we're learning from this new *Cosmos* series is that Tyson is "like that" after all—i.e., he's an atheist activist who is willing to rewrite history to suit his materialistic narrative, and intends to use *Cosmos* as a vehicle to promote the message.

48. The Hermetic Fellowship defines Hermeticism as "an ancient spiritual, philosophical, and magical tradition. It is a path of spiritual growth. Hermeticism takes its name from the God Hermes *Trismegistos* (Greek, "Thrice-Greatest Hermes"), a Graco-Egyptian form of the great Egyptian God of Wisdom and Magic, Thoth." See: http://www.hermeticfellowship.org/HFHermeticism.html.

49. Hank Campbell, "Cosmos: A Spacetime Odyssey—The Review," *Science 2.0* (7 March 2014), http://www.science20.com/science_20/blog/cosmos_spacetime_odyssey_review-131240.

50. Ibid.

7.

SLIPSHOD HISTORY

EVEN THE NATIONAL CENTER FOR SCIENCE
EDUCATION DECRIES "ANTI-RELIGIOUS BIAS" &
"SLIPSHOD HISTORY OF SCIENCE" IN *COSMOS*

Casey Luskin

WILLIAM DEMBSKI ONCE OBSERVED: "OUR CRITICS HAVE, IN EF-fect, adopted a zero-concession policy toward intelligent design. According to this policy, absolutely nothing is to be conceded to intelligent design and its proponents. It is therefore futile to hope for concessions from critics."[51] We see Dembski's rule at work in statements from a staff member at the National Center for Science Education (NCSE) regarding Discovery Institute's critiques of *Cosmos*. More on that in a moment.

We have already pointed out the program's anti-religious and historically inaccurate account of Giordano Bruno. So before going on, I must praise the keen analysis by Peter Hess, over on the NCSE's blog, of *Cosmos*'s revisionist account of Bruno. Dr. Hess charges that *Cosmos*'s treatment of Bruno reveals "considerable slipshod history of science and a curiously anti-religious bias." He notes:

51. William A. Dembski, "Dealing with the Backlash Against Intelligent Design," *Design Inference* (14 April 2004), http://www.designinference.com/documents/2004.04. Backlash.htm.

Of course, in 2014 we don't burn people at the stake, and except for the most conservative voices, Christians don't cast about casually labeling any dissenting theological perspective as "heresy."[52]

Dr. Hess, who is Catholic, goes on to provide a detailed and insightful account of what really happened with Bruno, explaining that while Bruno's persecution was of course a terrible tragedy, the man "clearly was not a martyr for modern science." Dr. Hess charges that *Cosmos's* treatment of Bruno reveals "considerable slipshod history of science and a curiously anti-religious bias." He notes:

> It is troubling that *Cosmos*—as its only historical background—chose to portray a fallaciously interpreted version of the tragic story of Giordano Bruno. It is unfortunate that the writers uncritically repeated a false narrative about the history of science and religion, providing a public who are already confused about the relationship between these two endeavors with misinformation rather than an accurate and balanced account of a complex history of interaction. The Catholic Church was not monolithic in its approach to science throughout the early modern period. There were both reactionary and profoundly progressive elements within the church, and some of Galileo's most important supporters were themselves clerics. If Neil deGrasse Tyson wants to recount some history of science in *Cosmos*, his writers have a professional obligation to furnish him with a text that honestly tells the story with all the subtlety it deserves.[53]

Bravo, Dr. Hess. These are much the same points that I've made, as has Jay Richards, and others as well.

But other Darwin defenders—especially atheistic ones—seem to confirm Dembski's predicted "zero-concession policy," refusing to admit when their own camp makes a mistake.

For example, in a piece titled, "Watch out, '*Cosmos*'! The Holy Inquisition is not happy with you" at *Salon*, Andrew Leonard praises the series for promoting "scientific materialism," but mocks Jay Richards's critique, stating:

---

52. Peter Hess, "A Burning Obsession: *Cosmos* and its Metaphysical Baggage," *Science League of America* (14 March 2014), http://ncse.com/blog/2014/03/burning-obsession-cosmos-its-metaphysical-baggage-0015452.

53. Ibid.

The revisionist *Cosmos* critique concerning Bruno goes like this: He wasn't even really a scientist, and he was burned to death because of his theological heresies and not his belief in Copernican theory, (SO HE DESERVED IT!) and the main reason he showed up on *Cosmos* at all was because he was "the only one with even a passing association with a scientific controversy to be burned at the stake during this period of history."

Maybe it's just me, but reading between the lines of this piece I detected what seemed to be a tinge of regret that unbelievers can no longer be punished for straying from the Gospel with purging fire. Neil Tyson—watch your back![54]

Leonard seems to have missed the fact that Jay Richards (and I) explicitly condemned Bruno's persecution in our respective articles. Leonard may amuse himself by making outlandish insinuations about Discovery Institute wanting to burn unbelievers, but the persecution of scientists is no laughing matter. Pro-Darwin scientists may not be burning critics at the stake, but a lot of people have had their careers harmed because of intolerance towards scientific viewpoints that dissent from neo-Darwinian evolution.

Another snarky pop-culture blog, "Happy Nice Time People," likewise mocks those who pointed out *Cosmos*'s anti-religious tone, stating: "No secret atheist agenda there. Feel better? Probably not, but we don't care."[55] Indeed, really die-hard defenders of *Cosmos*'s message probably don't care that Neil deGrasse Tyson rewrote history to bolster his anti-religious narrative.

Nevertheless, these defenders of *Cosmos* are now being contradicted by the top defenders of Darwinian evolution at the NCSE, who agree with us that the first episode showed a troubling "anti-religious bias," offering a "slipshod history of science," and that it was wrong to focus so much of the first episode on the persecution of Giordano Bruno.

---

54. Andrew Leonard, "Watch out, 'Cosmos'! The Holy Inquisition is not happy with you," *Salon* (11 March 2014), http://www.salon.com/2014/03/11/watch_out_cosmos_the_holy_inquisition_is_not_happy_with_you/.

55. Lisa Needham, "Creationists Watch 'Cosmos,' Emit Billions and Billions Of Sad Words," *Happy Nice Time People* (11 March 2014), http://happynicetimepeople.com/creationists-watch-cosmos-write-billions-billions-words-made-mad/.

For my part, I am happy to concede that Peter Hess over at the NCSE, in his apt analysis of Giordano Bruno, gets a lot of things very right.

Episode 2:
Some of the Things
That Molecules Do

8.

UNGUIDED ANSWERS

"MINDLESS EVOLUTION" HAS ALL THE ANSWERS— IF YOU DON'T THINK ABOUT IT TOO DEEPLY

Casey Luskin

WITH MORE EYE-POPPING CGI AND SPLENDID NEW SCENES OF Neil deGrasse Tyson touring the solar system in his high-tech spaceship, *Cosmos* Episode 2 weighs in on some of life's most profound questions. Toward the end of the episode, Tyson honestly admits, "Nobody knows how life got started," and even says, "We're not afraid to admit what we don't know," since "the only shame is to pretend we know all the answers." However, that came off as a nervously inserted qualification since the rest of the episode had so vigorously argued that what Tyson calls the "transforming power" of "mindless evolution" or "unguided evolution" indeed has all the answers to how life evolved on Earth. Except, that is, for a few cases where evolution was guided by human breeders, through "artificial selection."

Episode 2 structures its argument much as Charles Darwin did in *The Origin of Species*. The opening scenes discuss how human breeders artificially selected many different dog breeds from wolf-like ancestors, including many popular breeds that "were created in only the last few centuries." The argument is simple—and it's the same type of argument that Darwin made: "If artificial selection can work such profound changes in only 10 or 15 thousand years, what can natural selection do operating over billions of years?" The answer, Tyson tells us, is "all the beauty and diversity of life."

In other words, Tyson wants you to believe that natural selection provides all the answers for everything since life arose. Just as he did in Episode 1, Tyson has overstated his case. The great evolutionary biologist Ernst Mayr explains precisely why Tyson is wrong:

> Some enthusiasts have claimed that natural selection can do anything. This is not true. Even though "natural selection is daily and hourly scrutinizing, throughout the world, every variation even the slightest," as Darwin (1859:84) has stated, it is nevertheless evident that there are definite limits to the effectiveness of selection.[56]

Aside from the fact that artificial selection involves intelligent agents rather than unguided processes, Mayr makes one of the most important points in the context of artificial selection of dogs, for human breeders consistently hit limits in just how far they can breed dogs. The textbook *Explore Evolution: The Arguments For and Against Neo-Darwinism* explains:

> Intense programs of breeding (and inbreeding) frequently increase the organism's susceptibility to disease, and often concentrate defective traits. Breeders working with English bulldogs have strived to produce dogs with large heads. They have succeeded. These bulldogs now have such enormous heads that puppies sometimes have to be delivered by a Cesarean section. Newfoundlands and Great Danes are both bred for large size. They now have bodies too large for their hearts and can suddenly drop dead from cardiac arrest. Many Great Danes develop bone cancer, as well. Breeders have tried to maximize the sloping appearance of a German Shepherd's hind legs. As a result, many German Shepherds develop hip dysplasia, a crippling condition that makes it hard for them to walk. When breeders try to force a species beyond its limits, they often create more defects than desirable traits. These defects impose limits on the amount of change that breeders can ultimately produce.

Darwin's theory states that the unguided force of natural selection is supposed to be able to do what the intelligent breeder can do. But even a process of careful, intentional selection encounters limits that neither time nor the efforts of human breeders can overcome. Consequently, critics argue that by the logic of Darwin's own analogy, the power of natural selection is also limited.

---

56. Ernst Mayr, *What Evolution Is* (New York: Basic Books, 2001), 140.

Darwin's theory requires that species exhibit a tremendous elasticity—or capacity to change. Critics point out that this is not what the evidence from breeding experiments shows.[57]

These aren't just talking points from Darwin-critics. The same is heard from leading evolutionary biologists who say inconvenient things that *Cosmos* was content to ignore:

> The following are three major areas of misconception among the Neo-Darwinists.... Artificial selection on quantitative traits was taken as a model of the evolutionary process. It was easily shown, in agriculture or in the laboratory, that populations of most organisms contain sufficient additive genetic variance to obtain a response to selection on quantitative traits, such as measures of body size or increased yield of agriculturally valuable products such as milk in dairy cattle or grain size in food plants. Generalizing from this experience, it was assumed that natural populations are endowed with essentially unlimited additive genetic variance, implying that any sort of selection imposed by environmental changes will encounter abundant genetic variation on which to act. Moreover, this model was extended to evolutionary time as well as ecological time. This way of thinking ignored the substantial evidence from selection experiments that the response to selection on any trait essentially comes to a halt after a number of generations as the genetic variance for the trait in question is depleted; thereafter, further progress depends on the introduction of new variants either through outcrossing or new mutations (Falconer, 1981).[58]

Ernst Mayr concurs, citing "[t]he limited potential of the genotype" which shows "severe limits to further evolution" and explaining:

> The existing genetic organization of an animal or plant sets severe limits to its further evolution. As Weismann expressed it, no bird can ever evolve into a mammal, nor a beetle into a butterfly. Amphibians have been unable to develop a lineage that is successful in salt water. We marvel at the fact that mammals have been able to develop flight (bats) and aquatic adaptation (whales and seals), but there are many other ecological niches that mammals have been unable to occupy. There are, for instance, severe limits on size, and no amount of selection has allowed mammals to become

---

57. Stephen C. Meyer et al, *Explore Evolution: The Arguments For and Against Neo-Darwinism* (Malvern, Hill House, 2007), 91. See also: http://www.exploreevolution.com/.

58. Austin L. Hughes, "Looking for Darwin in all the wrong places: the misguided quest for positive selection at the nucleotide sequence level," *Heredity* 99 (2007), 364–373.

smaller than a pygmy shrew and the bumblebee bat, or allow flying birds to grow beyond a limiting weight.[59]

We cannot simply assert that evolution can do just "anything" or "all" we want it to—there are both genetic and physiological limits to how far breeders can change organisms. If we are to take artificial selection as an analogy for what can happen in the wild, shouldn't this suggest there are also limits to evolution?

I'm sure Tyson would reply that we can overcome genetic barriers to further evolution through mutations, which provide new raw materials for evolution to act upon. According to Tyson, mutations are "entirely random" and can cause changes like transforming a bear's fur color from brown to "white," as in "polar bears," giving it a camouflage advantage in a snowy environment. (Technically, *Cosmos* got this small detail wrong, since the hairs of polar bears are transparent, not white.[60]) "No breeder gathered these changes," he tells us, since "the environment itself selects them."

Fair enough. While we might disagree with Tyson that natural selection is "the most revolutionary concept in the history of science," no intelligent-design proponent denies that natural selection is an important idea that can explain many things. Changing the color of a bear's coat from brown to white is probably one of them—it's a small-scale, micro-evolutionary change. The difference between ID proponents and evolutionists like Tyson is that ID proponents acknowledge that natural selection is a real force in nature, but we don't just unconditionally grant it the power to do all things. Instead, we test forces like natural selection and find that there are limits to the amount of change it can effect in populations.

For example, after saying the "tree of life" (more on that shortly) is "three and a half billion years old," Tyson just asserts that this provides "plenty of time" for the evolution of life's vast complexity. But is this assertion true?

Tyson's Main Argument

TYSON'S MAIN ARGUMENT THAT SELECTION AND MUTATION CAN EVOLVE anything focuses on the evolution of the eye. Here, he attacks intelligent

---

59. Mayr, *What Evolution Is*, 140.
60. *Everyday Mysteries*, http://www.loc.gov/rr/scitech/mysteries/polarbear.html.

design by name, noting that some have argued that life "must be the work of an intelligent designer" that "created each of these species separately." I've never heard of an ID proponent who requires that every single species was created separately, so that's a straw man. Tyson calls the human eye a "masterpiece" of complexity and claims it "poses no challenge to evolution by natural selection." But do we really know this is true?

Darwinian evolution tends to work fine when one small change or mutation provides a selective advantage, or as Darwin put it, when an organ can evolve via "numerous, successive, slight modifications." If a structure cannot evolve via "numerous, successive, slight modifications," Darwin said, his theory "would absolutely break down." Evolutionary biologist Jerry Coyne essentially concurs: "It is indeed true that natural selection cannot build any feature in which intermediate steps do not confer a net benefit on the organism."[61] So are there structures that would require multiple steps to provide an advantage, where intermediate steps might not confer a net benefit on the organism? If you listen to Tyson's argument carefully, I think he let slip that there are.

In his account of the evolution of the eye, Tyson says that "a microscopic copying error" gave a protein the ability to be sensitive to light. He doesn't explain how that happened. Indeed, Sean B. Carroll cautions us to "not be fooled" by the "simple construction and appearance" of supposedly simple light-sensitive eyes, since they "are built with and use many of the ingredients used in fancier eyes."[62] Tyson doesn't worry about explaining how any of those complex ingredients arose at the biochemical level. More interesting is what Tyson says next: "Another mutation caused it [a bacterium with the light-sensitive protein] to flee intense light."

This raises an interesting question: It's nice to have a light-sensitive protein, but unless the sensitivity to light is linked to some behavioral response, then how would the sensitivity provide any advantage? Only once a behavioral response also evolved—say, to turn towards or away from the light—can the light-sensitive protein provide an advantage. So if a

---

61. Jerry Coyne, "The Great Mutator," *The New Republic* (June 14, 2007).
62. Sean B. Carroll, *The Making of the Fittest: DNA and the Ultimate Forensic Record of Evolution*, (New York: W. W. Norton, 2006), 197.

light-sensitive protein evolved, why did it persist until the behavioral response evolved as well? There's no good answer to that question, because vision is fundamentally a multi-component and thus a multi-mutation feature. Multiple components—both visual apparatus and the encoded behavioral response—are necessary for vision to provide an advantage. It's likely that these components would require many mutations. Thus, we have a trait where an intermediate stage—say, a light-sensitive protein all by itself—would not confer a net advantage on the organism. This is where Darwinian evolution tends to get stuck.

Indeed, ID research is finding that there are many traits that require many mutations before providing an advantage. For starters, protein scientist Douglas Axe has published mutational sensitivity tests on enzymes in the *Journal of Molecular Biology* and found that functional protein sequences may be as rare as 1 in 10^{77}. That extreme rarity makes it highly unlikely that chance mutations alone could find the rare amino acid sequences that yield functional proteins.[63]

According to Axe's research, many specific amino acids must be present in just the right sequence in order to yield enzyme functionality. This suggests many mutations must occur before an enzyme can function and provide any functional advantage. The waiting time for these mutations to arise would be enormous.

I'm sure the producers of *Cosmos* would reply that the "billions and billions" of years of evolution provide "plenty of time" even for such unlikely events. But unless we test this claim, it's a naïve response. In 2010, Axe investigated how many mutations could arise in a multi-mutation feature given the entire history of the Earth. He published population genetics calculations indicating that even when we grant generous assumptions favoring a Darwinian process, molecular adaptations requiring more than

---

63. Douglas Axe, "Estimating the Prevalence of Protein Sequences Adopting Functional Enzyme Folds," *Journal of Molecular Biology* 341 (2004), 1295–1315; Douglas Axe, "Extreme Functional Sensitivity to Conservative Amino Acid Changes on Enzyme Exteriors," *Journal of Molecular Biology* 301 (2000), 585–595.

six mutations before yielding any advantage would be extremely unlikely to arise in the 4.5 billion year history of our planet.[64]

The following year, Axe published research with developmental biologist Ann Gauger describing the results of their experiments seeking to convert one bacterial enzyme into another closely related enzyme—one in the same gene family! That is the kind of conversion that evolutionists claim can easily happen. For this case they found that the conversion would require a minimum of at least seven simultaneous changes,[65] exceeding the six-mutation-limit that Axe had previously established as a boundary of what Darwinian evolution is likely to accomplish in bacteria. Because this conversion is thought to be relatively simple, it suggests that converting one similar type of protein into another by "mindless evolution" might be highly unlikely.

In other experiments led by Gauger and biologist Ralph Seelke of the University of Wisconsin, Superior, their research team broke a gene in the bacterium *E. coli* required for synthesizing the amino acid tryptophan. When the bacteria's genome was broken in just one place, random mutations were capable of "fixing" the gene. But even when only two mutations were required to restore function, Darwinian evolution got stuck, apparently unable to restore full function.[66] Again, "mindless evolution" couldn't overcome the need to produce multi-mutation features—those that require multiple mutations before providing an advantage.

Theoretical research into population genetics corroborates these empirical findings. Michael Behe and David Snoke have performed computer simulations and theoretical calculations showing that the Darwinian evolution of a functional bond between two proteins would be highly unlikely to occur in populations of multicellular organisms under reasonable evolu-

---

64. Douglas Axe, "The Limits of Complex Adaptation: An Analysis Based on a Simple Model of Structured Bacterial Populations," *BIO-Complexity* 2010 no. 4 (2010), 1–10.

65. Ann Gauger and Douglas Axe, "The Evolutionary Accessibility of New Enzyme Functions: A Case Study from the Biotin Pathway," *BIO-Complexity* 2011 no. 1 (2011), 1–17.

66. Ann Gauger, Stephanie Ebnet, Pamela F. Fahey, and Ralph Seelke, "Reductive Evolution Can Prevent Populations from Taking Simple Adaptive Paths to High Fitness," *BIO-Complexity* 2010 no. 2 (2010), 1–9.

tionary timescales when it required multiple mutations before functioning. They published research in *Protein Science* that found:

> The fact that very large population sizes—10^9 or greater—are required to build even a minimal MR feature requiring two nucleotide alterations within 10^8 generations by the processes described in our model, and that enormous population sizes are required for more complex features or shorter times, seems to indicate that the mechanism of gene duplication and point mutation alone would be ineffective, at least for multicellular diploid species, because few multicellular species reach the required population sizes.[67]

In other words, in multicellular species, Darwinian evolution would be unlikely to produce features requiring more than just two mutations before providing any advantage on any reasonable timescale or population size.

In 2008, Behe's critics Rick Durrett and Deena Schmidt sought to refute him in the journal *Genetics* with a paper titled "Waiting for Two Mutations: With Applications to Regulatory Sequence Evolution and the Limits of Darwinian Evolution." But Durrett and Schmidt found that to obtain only two specific mutations via Darwinian evolution "for humans with a much smaller effective population size, this type of change would take >100 million years." The critics admitted this was "very unlikely to occur on a reasonable timescale."[68]

Plenty of Time

WHAT DOES THIS ALL MEAN? FIRST, IT MEANS COSMOS IS WRONG TO ASsert we know that there is "plenty of time" for the "mindless evolution" of complex structures to take place. Both theoretical and empirical research suggest there are very good reasons why producing many of the new proteins and enzymes entailed by eye-evolution, and probably many other evolutionary pathways, would require the generation of multi-mutation features that could not arise via "mindless evolution" in the 3.5 billion year

---

67. Michael Behe and David Snoke, "Simulating Evolution by Gene Duplication of Protein Features that Require Multiple Amino Acid Residues," *Protein Science* 13 (2004), 2651–2664.

68. Rick Durrett and Deena Schmidt, "Waiting for Two Mutations: With Applications to Regulatory Sequence Evolution and the Limits of Darwinian Evolution," *Genetics* 180 (November 2008), 1501–1509.

history of life on Earth. Second, it means *Cosmos* is pretending to have all the answers about how life evolved, when in fact it doesn't. And third, as David Berlinski has pointed out, it means that evolutionary biologists are very far away from explaining the evolution of the eye.[69]

The second episode of *Cosmos* showcased quite a lot of evolutionary apologetics. I mean attempts to persuade people of both evolutionary scientific views and larger materialistic evolutionary beliefs, not just by the force of the evidence, but by rhetoric and emotion, and especially by leaving out important contrary arguments and evidence. This episode focused its evolutionary apologetics on the tree of life.

Tyson states that we have an "understandable human need" to think that we're special, and thus "a central premise of traditional belief is that we were created separately from the other animals." If you believe that, then you should know that it's Neil deGrasse Tyson's intention to talk you out of that "traditional belief," and he's going to use beautiful animation to do it, while ignoring explanations like "common design" and otherwise misstating the evidence.

This episode shows a beautifully animated "tree of life," saying "science reveals that all life on Earth is one," and that "accepting our kinship with other animals" is "solid science." But it's not enough for Tyson if you just accept those evolutionary scientific views. The main message here is that humans aren't special, since we are just "one tiny branch among countless millions."

In case you think there's room for reasonable intellectual doubt, Tyson compares evolution to gravity, casting Darwinian evolution as an undeniable "scientific fact."

Perhaps common ancestry is a fact. But what is Tyson's evidence for it? It's this: similarities in DNA sequences between humans and other species. The episode portrays similar DNA sequences between humans and other species—butterflies, wolves, mushrooms, sharks, birds, trees, and even one-celled organisms—and says that because "we and other species are almost identical" in some core metabolic genes, "the DNA doesn't lie"

---

69. David Berlinski, "The Vampire's Heart," http://www.discovery.org/f/1061.

and we are "long-lost cousins" with all these other organisms. With evolutionary apologetics in full force, Tyson even says this realization offers a "spiritual experience"—a nice bit of "woo," included presumably to help appeal to the masses.

The Tree of Life in Tatters

SPIRITUAL OR NOT, IS IT TRUE THAT THERE'S A GRAND "TREE OF LIFE" showing how we're related to all other organisms?

A 2009 article in *New Scientist* concluded that the tree of life "lies in tatters, torn to pieces by an onslaught of negative evidence." Why? Because one gene yields one version of the tree of life, while another gene gives another sharply conflicting version of the tree. The article explained what's going on in this field:

> "For a long time the holy grail was to build a tree of life," says Eric Bapteste, an evolutionary biologist at the Pierre and Marie Curie University in Paris, France. A few years ago it looked as though the grail was within reach. But today the project lies in tatters, torn to pieces by an onslaught of negative evidence. Many biologists now argue that the tree concept is obsolete and needs to be discarded. "We have no evidence at all that the tree of life is a reality," says Bapteste. That bombshell has even persuaded some that our fundamental view of biology needs to change.[70]

According to the article:

> The problems began in the early 1990s when it became possible to sequence actual bacterial and archaeal genes rather than just RNA. Everybody expected these DNA sequences to confirm the RNA tree, and sometimes they did but, crucially, sometimes they did not. RNA, for example, might suggest that species A was more closely related to species B than species C, but a tree made from DNA would suggest the reverse.[71]

The problem is rampant in systematics today. An article in *Nature* reported that "disparities between molecular and morphological trees" lead to "evolution wars" because "[e]volutionary trees constructed by studying biological molecules often don't resemble those drawn up from morpholo-

---

70. Graham Lawton, "Why Darwin was wrong about the tree of life," *New Scientist* 2692 (January 21, 2009), 34–39.
71. Ibid.

gy."[72] Another *Nature* paper reported that newly discovered genes "are tearing apart traditional ideas about the animal family tree" since they "give a totally different tree from what everyone else wants."[73]

So severe are the problems that a 2013 paper in *Trends in Genetics* reported "the more we learn about genomes the less tree-like we find their evolutionary history to be,"[74] and a 2012 paper in *Annual Review of Genetics* proposed "life might indeed have multiple origins."[75]

Don't expect Neil deGrasse Tyson and *Cosmos* to disclose to viewers that there are problems with reconstructing a grand "tree of life." They need to maintain the pretense that "mindless evolution" provides all the answers.

If not by "mindless" or "unguided" evolution and common ancestry, how can we explain the fact that genes in different organisms are so similar? Though Neil deGrasse Tyson never mentions it, a fully viable explanation or these functional genetic similarities is common design.

Intelligent agents often re-use functional components in different designs, which means common design is an equally good explanation for the very data—similar functional genes across different species—that Tyson cites in favor of common ancestry. As Paul Nelson and Jonathan Wells explain:

> An intelligent cause may reuse or redeploy the same module in different systems, without there necessarily being any material or physical connection between those systems. Even more simply, intelligent causes can generate identical patterns independently.[76]

Likewise, in their book *Intelligent Design Uncensored*, William Dembski and Jonathan Witt explain:

72. Trisha Gura, "Bones, Molecules or Both?," *Nature* 406 (July 20, 2000), 230–233.

73. Elie Dolgin, "Rewriting Evolution," *Nature* 486 (28 June 2012), 460–462.

74. Bapteste et al., "Networks: expanding evolutionary thinking," *Trends in Genetics* 29 (2013), 439–41.

75. Michael Syvanen, "Evolutionary Implications of Horizontal Gene Transfer," *Annual Review of Genetics* 46 (2012), 339–56.

76. Paul Nelson and Jonathan Wells, "Homology in Biology," in *Darwinism, Design, and Public Education*, edited by John A. Campbell and Stephen C. Meyer (East Lansing: Michigan State University Press, 2003), 316.

According to this argument, the Darwinian principle of common ancestry predicts such common features, vindicating the theory of evolution.

One problem with this line of argument is that people recognized common features long before Darwin, and they attributed them to common design. Just as we find certain features cropping up again and again in the realm of human technology (e.g., wheels and axles on wagons, buggies and cars) so too we can expect an intelligent designer to reuse good design ideas in a variety of situations where they work.[77]

Thus, common design is a possible explanation for why two taxa can have highly similar functional genetic sequences. After all, designers regularly re-use parts, programs, or components that work in different designs. As another example, engineers use wheels on both cars and airplanes, or technology designers put keyboards on both computers and cell-phones. Or software designers re-use subroutines in different software programs.

But common designers aren't always obligated to design according to a nested hierarchy. So when we find re-use of functional components in a pattern that doesn't match a nested hierarchy, we might look to common design. That's exactly what we have here: similar genes being re-used in different organisms, but in a pattern that doesn't match the "tree-like" distribution predicted by Darwinian theory. Unfortunately, Neil deGrasse Tyson doesn't inform his viewers of any of this.

In this second episode of *Cosmos*, Tyson and his co-creators hoped to convince viewers that intelligent design is wrong, but simply by discussing the complexity of biology, they couldn't help but expose viewers to the evidence for design in nature.

When *Cosmos* Episode 2 showed brilliant animations of walking kinesin motors, and discussed the fact that DNA is a "molecular machine" that is "written in a language that all life can read," it unwittingly showed that intelligent design is a viable explanation. After all, what is the sole known cause that produces languages and machines? That one singular cause is, of course, intelligence. Even when you try to disregard the evidence for design in nature, it nevertheless speaks for itself.

---

77. William A. Dembski and Jonathan Witt, *Intelligent Design Uncensored: An Easy-to-Understand Guide to the Controversy* (Downers Grove: InterVarsity Press, 2010), 85.

Postscript

IN A REBUTTAL FILLED WITH *AD HOMINEM* ATTACKS, JOURNALIST CHRIS Mooney attempted to respond to this article by claiming that "science deniers" are "freaking out" over *Cosmos*.[78] His one substantive rebuttal is that the "tree of life" is doing just fine because of the "Open Tree of Life project, which plans to produce 'the first online, comprehensive first-draft tree of all 1.8 million named species, accessible to both the public and scientific communities.'" I'm sure that's a worthwhile project, but Mooney's comments don't address the fact that a "treelike pattern" is fundamentally incompatible with much of the data being discovered by molecular biology.

The conundrum that folks working with the Open Tree of Life project will face is this: Which tree is the real tree of life? They'll find that one gene gives you one version of the tree of life, and another gene gives an entirely different, conflicting version of the tree of life. The genetic data does not paint a consistent picture of common ancestry.

Mooney wants his readers to think these are isolated problems, since the attempt to "reconstruct every last evolutionary relationship may still be an open scientific question, but the idea of common ancestry, the core of evolution (represented conceptually by a tree of life), is not."

Actually, conflicts in the tree of life are rampant. As a 2012 paper stated, "Phylogenetic conflict is common, and frequently the norm rather than the exception," and "Phylogenetic conflict has become a more acute problem with the advent of genomescale data sets."[79] Or, as Michael Syvanen stated in the *New Scientist* article quoted above, "We've just annihilated the tree of life. It's not a tree any more, it's a different topology entirely." That seems to suggest scientists contributing to the "Open Tree of Life project" may be in for far greater difficulties than Mooney is letting on.

---

78. Chris Mooney, "Science Deniers Are Freaking Out About 'Cosmos,'" *Mother Jones* (17 March 2014), http://www.motherjones.com/blue-marble/2014/03/science-deniers-cosmos-neil-tyson.

79. Liliana M. Dávalos, Andrea L. Cirranello, Jonathan H. Geisler, and Nancy B. Simmons, "Understanding Phylogenetic Incongruence: Lessons from Phyllostomid Bats," *Biological Reviews of the Cambridge Philosophical Society* 87 (2012), 991–1024.

9.

BURNED AT THE RATINGS

IT'S A SHAME, REALLY, THAT COSMOS IS A "RATINGS DISASTER"

Evolution News & Views

IT'S TOO BAD THAT BY THE SECOND EPISODE HAD AIRED, COSMOS turned into a "ratings disaster," as a headline on *Drudge* summarized.[80] Fox spent a lot of money on this slick pop science excursion, intended to counter evolution skeptics and advocates of intelligent design. In Episode 2, Neil deGrasse Tyson took aim at the view that "Living things are just too intricate... to be the result of unguided evolution." This, of course, grossly distorts and simplifies what ID theorists say, but never mind.

Despite its many flaws, had the series really taken off as Fox hoped, that would have been a ripe opportunity to put the evolution debate before a very wide viewership. So the missed opportunity is regrettable.

However, we kept watching, and telling readers of *Evolution News & Views* what we thought. That sent Darwinists into fits. *Salon* staff writer Andrew Leonard, for one, didn't just insinuate, but all but stated outright that for Tyson's historical revisionism on Giordano Bruno, our mild-mannered colleague Jay Richards, who commented in Chapter 5, would like to see Neil Tyson burned at the stake! Writes the unsubtle Mr. Leonard:

---

80. "TV Ratings Sunday," *Zap2it*, (17 March 2014), http://tvbythenumbers.zap2it.
    com/2014/03/17/tv-ratings-sunday-resurrection-the-mentalist-revenge-down-believe-
    tumbles-mediocre-premiere-for-crisis/245084/.

Maybe it's just me, but reading between the lines of this piece I detected what seemed to be a tinge of regret that unbelievers can no longer be punished for straying from the Gospel with purging fire.[81]

A few comments on Twitter by Jay produced a long, sputtering outpouring of a post from biologist PZ Myers,[82] the briefly "happy" but now once again thoroughly dyspeptic atheist. Myers must have a light teaching load. There's nothing of substance in PZ's post, for all its prolixity, thus nothing to respond to.

Confusing intelligent design advocates with creationists and the "Christian Right," in typical fashion, writer Dan Arel at the *Huffington Post* exulted about how *Cosmos* makes us cower like "a frightened, cornered animal that knows it is about to die."[83] In Arel's, view, "the creationist lobby will simply stop at nothing to protect the industry that they have created," since "*Cosmos* frightens [us] more than anything has in a very long time."

Keep on dreaming, Mr. Arel. Why anyone would be terrified by what we aptly called "this big fat, lumbering Christmas turkey"[84] is a good question. Dan Arel is, like PZ Myers, long on excessive rhetoric and short on substance. Responding to a tweet from Jay Richards—"On eye evolution, the #Cosmos editors again failed to do a Google search"—Arel ventured:

> Tyson touched on how many species have evolved an eye, but did leave out the fact that there are over 40 known independent eye evolutions, something that very clearly discredits any intelligent design.

But this is just the lamest comment ever. The reason evolutionary biologists believe in "40 known independent eye evolutions" isn't because they've reconstructed those evolutionary pathways, but because eyes don't assume

81. Andrew Leonard, "Watch out, 'Cosmos'! The Holy Inquisition is not happy with you," *Salon* (11 March 2014), http://www.salon.com/2014/03/11/watch_out_cosmos_the_holy_inquisition_is_not_happy_with_you/.

82. PZ Meyers, "A Discovery Institute hack watches *Cosmos*," *ScienceBlogs* (18 March 2014), http://scienceblogs.com/pharyngula/2014/03/18/a-discovery-institute-hack-watches-cosmos/.

83. Dan Arel, "*Cosmos* Squashes Creationism Under the Weight of Evidence," *Huffington Post* (17 March 2014), http://www.huffingtonpost.com/dan-arel/cosmos-creationism_b_4976982.html.

84. David Klinghoffer, "The Second Episode of *Cosmos* with Neil deGrasse Tyson: Quick Reaction," *Evolution News & Views* (16 March 2014), http://www.evolutionnews.org/2014/03/the_second_epis083321.html.

a treelike pattern on the famous Darwinian "tree of life." Darwinists are accordingly forced, again and again, to invoke convergent "independent" evolution of eyes to explain why eyes are distributed in such a non-tree-like fashion.

This is hardly evidence against ID. In fact the appearance of eyes within widely disparate groups speaks eloquently of common design. Eyes are a problem all right—for Darwinism.

Had *Cosmos* scored higher with viewers, it would have provided a convenient occasion for talking about such matters. It's for that reason that we were rooting for Dr. Tyson and were sad to see him stumble.

Episode 3:
When Knowledge
Conquered Fear

10.

A RELIGIOUS NEWTON

NOW IT'S NEWTON: IDEOLOGY CONTINUES TO TRUMP HISTORY IN COSMOS

Jay W. Richards

THE THIRD INSTALLMENT OF COSMOS IS, ON THE SURFACE, LESS evangelizing in its materialism than were the first two installments. The best, truest segments of the episode are those that stick closely to the evidence. I enjoyed the discussion of the Oort cloud and the narrative thread about Isaac Newton, Robert Hooke, and Edmund Halley.

Unfortunately, host Neil deGrasse Tyson and the *Cosmos* producers have enshrouded this basic science with the same materialist narrative we've come to expect. Pre-modern peoples universally see false patterns and portents in the heavens, and invariably see the irregular specter of comets as portents of doom. We get the stereotypical contrast between a "prescientific world ruled by fear"—signaled by a cartoon drawing of a malevolent figure wearing a bishop's miter—and the emergence of modern science, which finally delivered us from such obscurantism.

This way of framing the history of science, however, requires a great deal of distortion and misrepresentation, especially when it comes to the figure of Isaac Newton. With Newton, the *Cosmos* writers encountered a dilemma: Either ignore his frankly religious and theistic view of reality or misrepresent and compartmentalize it. They chose the latter course.

Newton, says Tyson, was a "a God-loving man, he was also a genius." The very construction of the sentence subtly suggests a contrast between "God-loving" and "genius." One wonders whether an earlier draft of the script said "a God loving man, but also a genius."

In any case, Newton's religious ideas, along with his abiding interest in alchemy, are safely quarantined from his scientific work. Newton, we learn, was "obsessed with finding hidden messages in the Bible" and apparently spent a great deal of time with this obsession. Nevertheless, his lifelong research in "alchemy and biblical chronology never led anywhere."

Contrast that with his great scientific work in the *Optics* and the *Principia*. When Newton was born, Tyson explains, people thought the solar system had been created by God. In the *Principia*, which Tyson describes as the "opening pages of modern science," Newton succeeds in replacing God, that master clockmaker, with… gravity.

This would have been surprising to Newton. And it would be surprising to any literate reader of the *Principia*, and especially the section known as the *General Scholium*. There, Newton argued from the very clockwork of the solar system to the activity and existence of a transcendent God.

It's true that Newton's "teleo-mechanistic" view of the universe differed from the Aristotelian understanding that had dominated the pre-Copernican cosmology. Explaining this, however, would have required philosophical acumen far beyond what we've seen in *Cosmos*. Still—and here's the main point—Newton's actual view, and his actual argument in the *Principia*, are so blatantly theistic that only the most willful blindness could miss it.

Here's how Newton explains the orderly movement of the planets around the sun in the *General Scholium*:

> But it is not to be conceived that mere mechanical causes could give birth to so many regular motions: since the Comets range over all parts of the heavens, in very eccentric orbits. For by that kind of motion they pass easily through the orbs of the Planets, and with great rapidity; and in their aphelions, where they move the slowest, and are detain'd the longest, they recede to the greatest distances from each other, and thence suffer the least disturbance from their mutual attractions. This most beautiful System of the Sun, Planets, and Comets, could only proceed from the counsel and dominion of an intelligent and powerful being. And if the fixed Stars are

the centers of other like systems, these being form'd by the like wise counsel, must be all subject to the dominion of One; especially since the light of the fixed Stars is of the same nature with the light of the Sun, and from every system light passes into all the other systems.[85]

In other words, the intricate clockwork of our solar system and any other star systems that may exist required a master clockmaker. Indeed, Newton goes much farther than modern intelligent design arguments. He argues that the astronomical evidence points inexorably to the transcendent and all-powerful God of theistic belief.

At the time, Newton's appeal to a gravitational force—which implied that objects influence other objects instantaneously from a distance—seemed like occultism to the followers of Descartes whose intuitions were more decidedly materialistic. If, like the Cartesians, you view reality as a bunch of little sticky balls bouncing around in the void, then Newton's idea of gravity—instantaneous action at a distance—seems downright spooky. It was just too anti-materialist.

One need not defend the details of Newton's argument to see the obvious: In "the opening pages of modern science" one of the greatest geniuses of the scientific revolution made an argument for intelligent design. That's pretty good company.

---

85. Isaac Newton, "General Scholium," *The Newton Project Canada*, http://newtonprojectca. files.wordpress.com/2013/06/newton-general-scholium-1729-english-text-by-motte-letter-size.pdf.

11.

Science and Christianity

Cosmos Scrubs Religion's Positive Influence from the History of the Scientific Revolution

Casey Luskin

WE LIVE IN INTERESTING TIMES. ON THE ONE HAND, WE'RE CONstantly assured that science and religion don't conflict. At the same time, we're told—sometimes by the same people—that religion hinders science. Perhaps this is to be expected. Materialists want to project a religion-friendly image because popular culture expects it, while at the same time they make arguments that they hope will ultimately erode religious belief. This requires a tricky balancing act, which is on vivid display in the third episode of *Cosmos*, "When Knowledge Conquered Fear."

Our host, Neil deGrasse Tyson, doesn't dare directly attack religion. Instead, he artfully portrays humanity in early history as "an abandoned baby on a doorstep" with no idea how we got here, and no idea "how to end our cosmic isolation." As a result, he says that we looked for signs, hoping to find some "special meaning" in our world, but he says whenever we think we've found something "sacred," then we "deceive ourselves and others." Tyson claims that the "pre-scientific world" (e.g., the medieval era when religion dominated Western society) was "filled with fear," but scientists like Newton and Halley ushered in a "permanent revolution" in human thought, which "set us free."

Isaac Newton

As Jay Richards observed in Chapter 10, Tyson calls Newton a "God-loving man" and "a genius." At the same time, Tyson tells viewers that Newton's religious studies "never led anywhere." Similarly, Tyson acknowledges that Newton's work suggested the universe was "the work of a master clockmaker," but he says that Newton's appealing to God is "the closing of a door. It doesn't lead to other questions." In other words, it was only when Newton *wasn't* doing religion, and was doing science, that he contributed anything positive. If anything, Newton's religion hindered scientific advance. Tyson's message is simple: The best way to do good science is to throw off the shackles of religion.

History, in contrast, tells a different story. Early scientists including Newton were inspired to their scientific research precisely because of their religious beliefs. Newton was a monotheist who believed in a loving, truthful, personal God who would create an orderly, intelligible universe that God wanted us to discover and enjoy. It was these theological beliefs that propelled Newton to study the laws of nature.

Of course Tyson tells viewers none of this. To promote his revisionist history, he had to ignore numerous prominent historians of religion and science. Don't take it from me—take it from them. The eminent John Hedley Brooke writes scornfully:

> The implications of scientific advance for Christian theology are often reduced to a plausible but simplistic formula: as natural phenomena, formerly explained by the will of a deity, were increasingly understood in mechanistic terms, increasingly brought within the domain of natural laws, so the belief in an active, caring Providence was eroded until the God of Abraham, Isaac, and Jacob became nothing more than a remote clockmaker.[86]

In fact, Brooke tells us, while religion and science have had disagreements over the years, Newton and his contemporaries were inspired by belief in God to do science in the first place:

---

86. John Hedley Brooke, "Science and Theology in the Enlightenment," in *Religion and Science: History, Method, Dialogue*, W. Mark Richardson and Wesley J. Wildman eds. (New York: Routledge, 1996), 7.

Any suggestion that what was revolutionary in 17th-century thought was the complete separation of science from theology would be disqualified by Newton himself, who once wrote that the study of natural philosophy included a consideration of divine attributes and of God's relationship with the world....

Robert Boyle and Isaac Newton saw the study of nature as a religious duty. A knowledge of God's power and wisdom could be inferred from the intelligence seemingly displayed in the designs of nature. Newton affirmed that the natural sciences had prospered only in monotheistic cultures... He believed the universality of his laws was grounded in the omnipresence of a single divine Will.... If he is made to symbolize the new canons of scientific rationality, then it cannot be said that the scientific revolution saw a separation of science from theology.[87]

Brooke continues: "For Newton, as for Boyle and Descartes, there were laws of nature only because there had been a Legislator."[88] These early scientists wanted to discover the laws that God had built into the natural world. They searched for those laws because they believed in a "Legislator." *Cosmos* again ignores, contradicts, and revises this history.

Lest you dismiss John Hedley Brooke, he is no religious apologist. He formerly taught at Oxford University and has served as president of the British Society for the History of Science and the International Society for Science and Religion.

Another eminent scholar of science and religion worth considering is Ian G. Barbour. He writes: "Newton himself believed that the world-machine was designed by an intelligent Creator and expressed God's purposes."[89] He explains just how profound an influence religion had in inspiring science in England during the crucial early stages of the scientific revolution:

> The English authors whom we would call scientists called themselves "natural philosophers" or "virtuosi." They were mainly from Anglican (Church of England) and Puritan (Calvinist) backgrounds. The charter of the Royal Society instructed its fellows to direct their studies "to the glory of God

87. Ibid, 8.
88. Ibid, 9.
89. Ian G. Barbour, *Religion and Science: Historical and Contemporary Issues* (San Francisco: HarperSanFrancisco, 1997), 18.

and the benefits of the human race." Robert Boyle (1627–1691) said that science is a *religious task*, "the disclosure of the admirable workmanship which God displayed in the universe." Newton believed the universe bespeaks an all-powerful Creator. Sprat, the historian of the Royal Society, considered science a valuable aid to religion....

The virtuosi identified themselves with the Christian tradition in which they were nourished, and many of them seem to have experienced a personal response of *reverence* and *awe* toward the marvels they beheld.... The sense of the grandeur and wisdom of God was evidently a very positive experience for many of them and not just an abstract intellectual formula or a concession to cultural respectability.[90]

It's not hard to understand how the founders of the world's most prestigious and long-lasting scientific society—who believed in a wise and powerful Creator—would then be inspired to investigate the "workmanship" of that Creator. In its Episode 3, *Cosmos* admits that early intellectual giants of science like Hooke, Newton, and Halley were a part of this same society, but *Cosmos* omits any mention of the important religious influences within that group. Religion and the founding of modern science went hand-in-hand, but *Cosmos* doesn't tell viewers *any of that*.

Johannes Kepler

ANOTHER GREAT FIGURE IN THE EARLY HISTORY OF MODERN SCIENCE, Johannes Kepler (1571–1630) wrote in *The Harmony of the World* (1618):

> The heavenly motions are nothing but a continuous song for several voices (perceived by the intellect, not the ear); a music which, through discordant tensions, through sincopes and cadenzas, as it were (as men employed them in imitation of natural discords) progresses towards certain predesigned, quasi six-voiced clausuras, and thereby sets landmarks in the immeasurable flow of time. It is, therefore, no longer surprising that man, in imitation of his creator, has at last discovered the art of figured song, which was unknown to the ancients. Man wanted to reproduce the continuity of cosmic time within a short hour, by an artful symphony for several voices, to obtain a sample test of the design of the Divine Creator in His works, and to partake of his joy by making music in imitation of God.[91]

---

90. Ibid., 19–20.
91. Richard P. Olenick and Tom M. Apostol, *The Mechanical Universe: Introduction to Mechanics and Heat* (Cambridge: Cambridge University Press, 1985), 553.

Western Science

IAN BARBOUR ASKS, "WHY WAS IT IN *WESTERN CIVILIZATION* ALONE, among all the cultures of the world, that science in its modern form developed?"[92] Because only the West had what Barbour says are the needed *"intellectual presuppositions* underlying the rise of science." Those presuppositions came directly from the Judeo-Christian worldview, which had uniquely permeated the West. Barbour explains:

> [T]he medieval legacy also included presuppositions about nature that were congenial to the scientific enterprise. First, the conviction of the *intelligibility of nature* contributed to the rational or theoretical component of science. The medieval scholastics.... combined the Greek view of the orderliness and regularity of the universe with the biblical view of God as Lawgiver. Monotheism implies the universality of order and coherence....
>
> Second, the doctrine of creation implies that *the details of nature can be known only by observing them.* For if the world is the product of God's free act, it did not have to be made as it was made, and we can understand it only by actual observation....
>
> Third, an *affirmative attitude towards nature* is dominant in the Bible. The goodness of the world is a corollary of the doctrine of creation.[93]

Barbour concludes: "many historians of science have acknowledged the importance of the Western religious tradition in molding assumptions about nature that were congenial to the scientific enterprise."[94]

Holmes Rolston III, Distinguished Professor of Philosophy at Colorado State University and winner of the Templeton Prize, likewise explains:

> Indeed, to turn the tables, it was monotheism that launched the coming of physical science, for it promised an intelligible world, sacred but disenchanted, a world with a blueprint, which was therefore open to the searches of the scientists. The great pioneers in physics—Newton, Galileo, Kepler, Copernicus—devoutly believed themselves called to find evidences of God in the physical world. Even Einstein, much later and in a different era, when puzzling over how space and time were made, used to ask himself how God would have arranged the matter. A universe of such beauty, an Earth given over to life and to culture—such phenomena imply

---

92. Barbour, *Religion and Science,* 27. Emphasis in original.
93. Ibid., 28.
94. Ibid., 29.

a transcending power adequate to account for these productive workings in the world.[95]

Yet another scholar, David C. Lindberg of the University of Wisconsin-Madison, former president of the U.S. History of Science Society, writes:

> There was no warfare between science and the church. The story of science and Christianity in the Middle Ages is not a story of suppression nor one of its polar opposite, support and encouragement. What we find is an interaction exhibiting all of the variety and complexity that we are familiar with in other realms of human endeavor: conflict, compromise, understanding, misunderstanding, accommodation, dialogue, alienation, the making of common cause, and the going of separate ways. Out of this complex interaction (rather than by repudiation of it) emerged the science of the Renaissance and the early-modern period.[96]

Or consider another authority, Discovery Institute Fellow Michael Keas, who explained in *Salvo Magazine*:

> Are Christianity and science at war with one another? Not according to leading historians. "The greatest myth in the history of science and religion holds that they have been in a state of constant conflict," wrote historian of science Ronald Numbers in 2009. Even though he and other historians of science have documented this conclusion thoroughly, many myths about the alleged warfare between science and religion continue to be promulgated in the popular literature and textbooks.
>
> The truth is that science and biblical religion have been friends for a long time. Judeo-Christian theology has contributed in a friendly manner to such science-promoting ideas as discoverable natural history, experimental inquiry, universal natural laws, mathematical physics, and investigative confidence that is balanced with humility. Christian institutions, especially since the medieval university, have often provided a supportive environment for scientific inquiry and instruction.
>
> Why have we forgotten most of the positive contributions of Christianity to the rise of modern science? This cultural amnesia is largely due to the influence of a number of anti-Christian myths about science and religion.

---

95. Holmes Rolston, III, *Science and Religion: A Critical Survey* (Harcourt Brace, 1987), 39.

96. David C. Lindberg, "Medieval Science and Religion," in *The History of Science and Religion in the Western Tradition: An Encyclopedia*, Gary B. Ferngren ed. (New York: Garland Publishing, 2000), 266.

These myths teach that science came of age in the victory of naturalism over Christianity.[97]

Dr. Keas goes on to debunk these myths. He explains in detail how Judeo-Christian theology provided the fertile ground that was necessary and natural for that development.

Cosmos Episode 3 tells us that Isaac Newton's religion was like "the closing of a door," and "never led anywhere" because belief in God "doesn't lead to other questions." This is historically wrong. A chorus of modern historians of science and religion observe that Judeo-Christian conceptions of God played an important, positive role in the rise of modern science. Not only that, but Judeo-Christian religion is crucial in answering a question that has long puzzled historians: *Why did modern science only arise in the West? Cosmos* whitewashes all of this, presenting instead a shallow revisionist formula: science good, religion bad. If the founders of modern science themselves thought in such simplistic terms, we never would have had science.

97. Michael Keas, "In the Beginning," *Salvo* (Fall 2013), http://salvomag.com/new/articles/salvo26-science-faith/in-the-beginning.php.

12.

NEWTON'S CRYPTOGRAM

THEY LEFT THIS OUT OF *COSMOS*: NEWTON SAW NATURE AS A "CRYPTOGRAM SET BY THE ALMIGHTY"

David Klinghoffer

As I was watching what Neil deGrasse Tyson does with Isaac Newton, I thought of John Maynard Keynes's summary of the relationship of Newton's science to his religious beliefs:

> He regarded the universe as a cryptogram set by the Almighty—just as he himself wrapt the discovery of the calculus in a cryptogram when he communicated with Leibniz. By pure thought, by concentration of mind, the riddle, he believed, would be revealed to the initiate.[98]

Keynes believed that this made Newton the "last of the magicians." Actually, it made him a great scientist. Yes, a "cryptogram set by the Almighty"—as a characterization of the puzzles offered by our own contemporary science, unsuccessfully whitewashed by a materialism that assures us it's got everything all figured out, I don't know of any descriptive phrase that's better.

---

98. John Maynard Keynes, "Newton, the Man" (London: Royal Society of London, 1946), http://phys.columbia.edu/~millis/3003Spring2014/Supplementary/John Maynard Keynes_ %22Newton, the Man%22.pdf.

EPISODE 4:
A SKY FULL OF GHOSTS

13.

FARADAY & MAXWELL

AS CHRISTIANS

FEWER ERRORS OF COMMISSION, MORE ERRORS OF OMISSION

Casey Luskin

A S THE FOURTH EPISODE OF COSMOS BEGAN, I WAS AGAIN ENJOYING Neil deGrasse Tyson's CGI-based tour of our universe, and thought, "This is what *Cosmos* should have been—pure science, spectacular visual effects, and no contrived potshots at religion. This is the perfect recipe to get the public interested in science!" Had *Cosmos's* first four episodes stuck to that science-focused recipe, I would have had very little negative to say. But they didn't. Knowing how *Cosmos* had already decided to "get ideological" in its other episodes, what's noteworthy about Episode 4 isn't just what it does say, but what it doesn't.

Episode 4 takes viewers on another tour of cosmic history, explaining what happened at various stages after the Big Bang, especially with regard to light. For the most part, it followed the "recipe" more closely than other episodes—which is a good thing—and again reflected what *Cosmos* should have been. Towards the end, there is a striking animated scene where Tyson's spaceship approaches, and then passes into, a black hole. Science-wise, it's a fascinating scene. But then Tyson says, "Maybe that's how our cosmos came to be," suggesting that our universe exists inside a black hole inside of another universe, which itself is a black whole inside of another universe, and so on, and so on. He admits that this idea is completely "speculative." But there's a not-so-hidden motivation here—he wants to lend credence to the "multiverse" hypothesis. Why is that?

Since this episode of *Cosmos* reviewed the development of the universe, it would have been an ideal opportunity to mention the fact that many prominent physicists believe the laws of the universe are finely tuned to allow for life to exist.[99] But there was no mention of this, nor was there any mention that Nobel laureate Charles Townes thinks this "fine-tuning" points to intelligent design:

> Intelligent design, as one sees it from a scientific point of view, seems to be quite real. This is a very special universe: it's remarkable that it came out just this way. If the laws of physics weren't just the way they are, we couldn't be here at all. The sun couldn't be there, the laws of gravity and nuclear laws and magnetic theory, quantum mechanics, and so on have to be just the way they are for us to be here.[100]

But of course the entire motive behind proposing the multiverse is to escape the improbability of our finely tuned universe. Charles Townes continues:

> Some scientists argue that "well, there's an enormous number of universes and each one is a little different. This one just happened to turn out right." Well, that's a postulate, and it's a pretty fantastic postulate—it assumes there really are an enormous number of universes and that the laws could be different for each of them. The other possibility is that ours was planned, and that's why it has come out so specially.

Cosmos decides to endorse something like the "fantastic postulate" and doesn't even mention the possibility that intelligent design is a viable way, according to some very prominent scientists, to explain cosmic fine-tuning.

Michael Faraday and James Maxwell

THERE ARE MORE OMISSIONS. IN THE PREVIOUS EPISODE, TYSON HAD INaccurately claimed that the religious beliefs of early founders of science like

99. Casey Luskin, "Roger Penrose on Cosmic Fine-Tuning," *Evolution News & Views* (12 April 2010), http://www.evolutionnews.org/2010/04/roger_penrose_on_cosmic_finetu033691.html.

100. Bonnie A. Powell, "Explore as much as we can," *UCBerkeleyNews* (17 June 2005), http://berkeley.edu/news/media/releases/2005/06/17_townes.shtml.

Isaac Newton didn't do anything to positively foster scientific discovery. This latest episode covers some additional giants from the annals of science, such as Michael Faraday and James Clerk Maxwell, but again omits any mention of their strong, positive religious influences. Ian Hutchinson, MIT Professor of Nuclear Science and Engineering, explains how Faraday and Maxwell were committed Christians:

> Two great British scientists dominate the intellectual landscape of electrical science, and indeed all of physics, in the nineteenth century, Michael Faraday and James Clerk Maxwell. It would be hard to imagine two more contrasting personalities.
>
> .... The contrasts between these men could be multiplied on and on. Yet they had one experience in common. Both were committed Christians.[101]

In another article, Hutchinson explains that Faraday's inspiration to discover intelligible laws of nature came from his "metaphysical presuppositions that directed his research":

> Faraday believed that in his scientific researches he was reading the book of nature, which pointed to its creator, and he delighted in it: "for the book of nature, which we have to read is written by the finger of God."
>
> One example of the influence of his theological perspective on his science is Faraday's preoccupation with nature's laws. "God has been pleased to work in his material creation by laws," he remarked, and "the Creator governs his material works by definite laws resulting from the forces impressed on matter." This is part of the designer's art: "How wonderful is to me the simplicity of nature when we rightly interpret her laws...."
>
> Another guiding principle of much of Faraday's thought finds its motivation in a conception of creation as a divinely planned economy. It is the principle of "conservation of force."[102]

Tyson notes that Faraday discovered these invisible fields of force, helping us explain how gravity works. But of course he doesn't say anything about how religion played a positive role in Faraday's scientific thinking.

---

101. Ian H. Hutchinson, "James Clerk Maxwell and the Christian Proposition," *MIT Independent Activities Period Seminar* (January 1998 & 2006), http://silas.psfc.mit.edu/Maxwell/maxwell.html.

102. Ian H. Hutchinson, "Michael Faraday: Scientist and Nonconformist," *MIT Independent Activities Period* (14 January 1996), http://silas.psfc.mit.edu/Faraday/.

Then, Tyson notes that James Clerk Maxwell "translated Faraday's idea about fields of electricity and magnetism into mathematical laws." Again, Tyson fails to mention Maxwell's deep religious faith. And again, Professor Hutchinson explains what *Cosmos* won't. He quotes Maxwell explaining how his Christian faith allows him to discover what's found in nature:

> Now my great plan, which was conceived of old… is to let nothing be willfully left unexamined. Nothing is to be holy ground consecrated to Stationary Faith, whether positive or negative. All fallow land is to be ploughed up and a regular system of rotation followed.… Never hide anything, be it weed or no, nor seem to wish it hidden.… Again I assert the Right of Trespass on any plot of Holy Ground which any man has set apart.… Now I am convinced that no one but a Christian can actually purge his land of these holy spots.… I do not say that no Christians have enclosed places of this sort. Many have a great deal, and every one has some. But there are extensive and important tracts in the territory of the Scoffer, the Pantheist, the Quietist, Formalist, Dogmatist, Sensualist, and the rest, which are openly and solemnly Tabooed.…
>
> Christianity—that is, the religion of the Bible—is the only scheme or form of belief which disavows any possessions on such a tenure. Here alone all is free. You may fly to the ends of the world and find no God but the Author of Salvation. You may search the Scriptures and not find a text to stop you in your explorations.…
>
> The Old Testament and the Mosaic Law and Judaism are commonly supposed to be "Tabooed" by the orthodox. Sceptics pretend to have read them, and have found certain witty objections… which too many of the orthodox unread admit, and shut up the subject as haunted. But a Candle is coming to drive out all Ghosts and Bugbears. Let us follow the light.[103]

What a striking quote to end a review of another episode of *Cosmos*. Throughout the series, Neil deGrasse Tyson repeatedly alludes to the rhetoric of Carl Sagan, casting science as the "light" that helps us understand our world, freeing us from the darkness of religion. Driven by a materialistic agenda, *Cosmos* shuts up as haunted the tabooed territory of religion's positive influence on the development of science.

103. Ian H. Hutchinson, "Michael Faraday: Scientist and Nonconformist," *MIT Independent Activities Period* (14 January 1996), http://silas.psfc.mit.edu/Faraday/.

Episode 5:
Hiding in the Light

14.

Mozi's Heaven

Neil Tyson Enlists a Chinese Philosopher in the Argument Against Faith—No, Make that *Fate*

David Klinghoffer

THE FIFTH INSTALLMENT OF *COSMOS*, "HIDING IN THE LIGHT," AGAIN beats the drum for its theme of science versus faith, with science cast in the recurring role of martyr to sinister, narrow-minded, probably religiously conservative forces.

Neil Tyson opens with an interesting depiction of a Chinese philosopher, Mo Tze, also spelled Mozi, presented as a forward-looking advocate of scientific reasoning, insisting on proof, "questioning authority," arguing "against blind obedience to ritual and authority," etc. He is credited with an early description of a camera obscura, relevant to the episode's theme of light and its properties.

One of Mozi's essays is said by Dr. Tyson to have been "Against Faith"—that's what it sounds like Tyson says, although I evidently misheard, as I would guess most viewers did. Mozi *did* argue against fate and fatalism. One of the essays attributed to his school is "Rejecting Fatalism," not "Faith."

After his death, bad guys are shown burning his works and burying his followers alive. Books or people being burned for trying to advance science is now a standard motif in *Cosmos*.

I had a suspicion that this is all a bit tendentious, and indeed the *Stanford Encyclopedia of Philosophy* gives a less propagandizing take.[104] Sure enough, I asked our friend and contributor Dr. Stephen Webb for his view. Was Mozi the standard-bearer of *Cosmos*-style materialism that Tyson seems to imply? Webb missed the episode, but replied:

> I've taught Chinese religion before, and everyone knows that Mozi was more religious than Confucius. Mozi actually appealed to Heaven (*Tian*) as an active agent in the world, which Confucius did not. He urged the state to encourage religious practices, another thing that Confucius didn't! The Communists liked him because he criticized Confucius's emphasis on tradition and the family. The Communists argued that his belief in ghosts and spirits was purely pragmatic, but no scholars today doubt the sincerity of his belief in the supernatural. His texts were suppressed, but the reason had to do ONLY with the victory of Confucianism and the fact that he was such an outspoken critic of Confucianism. Who writes these episodes anyway?

Yes, wouldn't it be fascinating to have been a fly on the wall at the storyboard meetings that produced these scripts?

Tyson also lectures us, "Science needs the light of free expression to flourish. It depends on the fearless questioning of authority, the open exchange of ideas." He's right, but about when would you say that scientific culture, as represented in academia and the media, decided that adherence to authority is actually the more "scientific" attitude? Somewhere after 1859, I guess.

Tyson is an endearing man, despite the faults of this series. I like the way he solicitously turns his voice up in a question mark when the grammar of the sentence doesn't call for it. This episode is mostly innocuous and otherwise very nicely done.

---

104. "Mohism," *Stanford Encyclopedia of Philosophy* (21 October 2002, revised 14 May 2014), http://plato.stanford.edu/entries/mohism/.

15.

MOZI'S MONOTHEISM

SHINING LIGHT ON THE LATEST ERRORS AND OMISSIONS IN *COSMOS*

Casey Luskin

T HE FIFTH INSTALLMENT OF *COSMOS* IS DESCRIBED ON THE PRO-
gram's website as an opportunity to "Discover the meanings of light
and enlightenment." The episode has some wonderful animations illustrat-
ing how electrons are bumped into higher energy levels when they absorb
light, and how they then emit light when they drop to a lower energy level.
It includes lucid explanations of how each element has a unique absorp-
tion and emission spectrum of light, which amazingly allows us to detect
the presence (or absence) of specific elements in stars that are light years
away simply by studying the spectrum of light the stars emit. There are also
keen comparisons of different types of electromagnetic radiation to the
"octaves" of sound in music—an effective audiovisual method of explaining
the electromagnetic spectrum. I know science teachers who would love to
use this sort of material in their classes—if only it weren't consistently re-
vising history and promoting inaccuracies to advocate a demonstrably false
materialistic narrative of science.

Early in the episode, Neil deGrasse Tyson discusses the ancient Chinese philosopher Mozi, whose ideas included "early stirrings of the scientific approach," as well as innovative political theories encouraging peace,
love, and egalitarian values. Perhaps unsurprisingly, Mozi's followers were
later persecuted by a government that wanted power.

As I watched the episode, my notes record that I heard Tyson say Mozi wrote a book titled *Against Faith*, as if Tyson intended to suggest that Mozi was some early anti-religious visionary. Later, after finishing the episode, I read David Klinghoffer's excellent response (Chapter 14) and was surprised to learn that the actual title was *Against Fate*—or according to the *Stanford Encyclopedia of Philosophy*, which describes it[105] as an essay rather than a book, "Rejecting Fatalism."[106]

So what did Tyson actually say in the episode? Is this another historical inaccuracy in *Cosmos*? I've listened to Tyson's words now multiple times, and here's my conclusion: It could be argued that he says Mozi's work was titled "Against Fate," but when you listen over and over again, it sounds a lot more like he's saying "Against Faith."

Certainly, the context has nothing to do with fate or fatalism, but it has everything to do with a supposed triumph of rational investigation over faith-based thinking. Our colleagues Jay Richards and Donald McLaughlin also heard "Against Faith." Donald adds that these programs are of course taped and meticulously edited, so it's not an unscripted live performance where verbal stumbles or ambiguities are unavoidable and difficult or impossible to correct. You can listen to the clip and decide for yourself.[107]

If Tyson was seeking a materialist sage to spark science in ancient China, then Mozi is the wrong guy. One statement by Tyson that is clear comes when he praises Mozi for promoting a philosophy "against blind obedience to ritual and authority," attempting to cast Mozi as some kind of a secular innovator. Once again, Tyson left out a crucial, inconvenient fact: *Mozi was a monotheist who scholars have recognized promoted a "Christian"-like view of God.* You might even call Mozi an apologist for a form of monotheistic religion in his day. As historian Klaus Schlichtmann puts it:

> Mozi advocated a monotheistic religion, in which God reigned as King in Heaven, a universalism based on principles of equality and justice, as well as the concept of "unbound (i.e., undifferentiated) love" (*jian'ai*), which was

---

105. "Mohism," *Stanford Encyclopedia of Philosophy* (21 October 2002, revised 14 May 2014), http://plato.stanford.edu/entries/mohism/.

106. "Mozi," Wikipedia, http://en.wikipedia.org/wiki/Mozi.

107. "Against Faith," *Evolution News & Views*, http://www.evolutionnews.org/againstfaith.mp3.

also said to be of "mutual utility," quite similar to the Christian idea in many ways.

The Chinese scholar and reformer Hu Shi (1891–1962) remarked in 1919 that Mozi was "probably the only Chinese who had founded a religion" and "possibly one of the greatest spirits China ever produced."[108]

Far from being against faith, Mozi founded a monotheistic religion where a supreme and loving God reigned over the Earth from heaven. No wonder he also promoted scientific methodologies. After all it was also a monotheistic culture—a Christian one—that gave birth to science in the West (see Chapter 11), where people believed in one God who reigned supreme over the universe and gave it intelligible, discoverable laws. Once again, we see that monotheistic religion is conducive to science and democratic values. *Cosmos* not only ignores this, but seeks to give the impression that religion and science stand opposed to each other.

One aspect of this episode that I really liked was Neil deGrasse Tyson's strong statements about the importance of intellectual freedom for a healthy science. He says "science needs the light of free expression to flourish," and notes that science "depends on the fearless questioning of authority," and requires "the open exchange of ideas." That's exactly right—bravo Dr. Tyson!

Unfortunately, Tyson stops short of asking whether scientists today have the academic freedom to question certain authorities or freely express certain views. So let's ask a question that *Cosmos* wouldn't: *Are scientists today free to express their views when they feel there are problems with authoritative paradigms, like modern evolutionary biology?* Don't ask me. Ask scientists and skeptics:

Keeping Dirty Laundry Hidden

First, there's Daniel Hillis, an MIT-trained scientist and engineer, who wrote in 1995:

> There's a feeling in biology that scientists should keep their dirty laundry hidden, because the religious right are always looking for any argument

---

108. Klaus Schlichtmann, *Japan in the World: Shidehara Kijuro, Pacifism, and the Abolition of War* (Lanham: Lexington Books, 2009), 12–13. Internal citations removed.

between evolutionists as support for their creationist theories. There's a strong school of thought that one should never question Darwin in public.[109]

High Implausibility as an Accident

TURNING TO A DIFFERENT FIELD, THERE'S THOMAS NAGEL, PROFESSOR of law at New York University and the author of *Mind and Cosmos* who says:

> It is *prima facie* highly implausible that life as we know it is the result of a sequence of physical accidents together with the mechanism of natural selection....
>
> My skepticism is not based on religious belief or on a belief in any definite alternative. It is just a belief that the available scientific evidence, in spite of the consensus of scientific opinion, does not in this matter rationally require us to subordinate the incredulity of common sense. This is especially true with regard to the origin of life....
>
> I realize that such doubts will strike many people as outrageous, but that is because almost everyone in our secular culture has been browbeaten into regarding the reductive research program as sacrosanct, on the ground that anything else would not be science....
>
> In thinking about these questions I have been stimulated by criticisms of the prevailing scientific world picture... by the defenders of intelligent design.... [T]he problems that these iconoclasts pose for the orthodox scientific consensus should be taken seriously. They do not deserve the scorn with which they are commonly met. It is manifestly unfair.[110]

Problems with Biologists' Dogma

IN THEIR 2003 BOOK, ACQUIRING GENOMES, LYNN MARGULIS AND Dorion Sagan write:

> Honest critics of the evolutionary way of thinking who have emphasized problems with biologists' dogma and their undefinable terms are often dismissed as if they were Christian fundamentalist zealots or racial bigots. But the part of this book's thesis that insists such terminology interferes

---

109. W. Daniel Hillis, "Introduction: The Emerging Third Culture," *Third Culture: Beyond the Scientific Revolution*, John Brockman, ed. (New York: Simon & Schuster, 1995), 26.

110. Thomas Nagel, *Mind and Cosmos: Why the Materialist Neo-Darwinian Conception of Nature Is Almost Certainly False* (New York: Oxford University Press, 2012), 6–7, 10.

with real science requires an open and thoughtful debate about the reality of the claims made by zoocentric evolutionists.[111]

No Satisfactory Explanation for Macroevolution

THEN THERE IS THIS REMARK FROM GÜNTER THEISSEN, A GENETICIST at the Freidrich Schiller University in Jena, Germany.

> It is dangerous to raise attention to the fact that there is no satisfying explanation for macroevolution. One easily becomes a target of orthodox evolutionary biology and a false friend of proponents of non-scientific concepts. According to the former we already know all the relevant principles that explain the complexity and diversity of life on earth; for the latter science and research will never be able to provide a conclusive explanation, simply because complex life does not have a natural origin.[112]

Neo-Darwinism Unquestioned

WRITING IN *WHAT DARWIN GOT WRONG*, A PHILOSOPHER, JERRY Fodor, and a cognitive scientist, Massimo Piattelli-Palmarini, note:

> We've been told by more than one of our colleagues that, even if Darwin was substantially wrong to claim that natural selection is the mechanism of evolution, nonetheless we shouldn't say so. Not, anyhow, in public. To do that is, however inadvertently, to align oneself with the Forces of Darkness, whose goal is to bring Science into disrepute....
>
> [N]eo-Darwinism is taken as axiomatic; it goes literally unquestioned. A view that looks to contradict it, either directly or by implication, is *ipso facto* rejected, however plausible it may otherwise seem. Entire departments, journals and research centres now work on this principle.[113]

Fear of Attacks

INDEED, WHILE SOME SCIENTISTS ARE FORCED TO CENSOR THEIR CRITIcisms of Darwinian theory out of fear of the establishment, a 2008 article in *Nature* on the Altenberg 16 conference explained that some others *will-*

---

111. Lynn Margulis and Dorion Sagan, *Acquiring Genomes: A Theory of the Origins of the Species* (New York: Basic Books, 2003), 29.

112. Günter Theißen, "The proper place of hopeful monsters in evolutionary biology," *Theory in Biosciences* 124 (2006), 349–369.

113. Jerry Fodor and Massimo Piattelli-Palmarini, *What Darwin Got Wrong* (New York: Farrar, Straus and Giroux, 2010), xx, xvi.

ingly self-censor their own criticisms so as to avoid "handing ammunition" to "creationists":

> [T]here was no sense at Altenberg of a desire to attack evolutionary theory from the left. Quite the reverse—the dominant political concern was a fear of attack from fundamentalists. As Gould discovered, creationists seize on any hint of splits in evolutionary theory or dissatisfaction with Darwinism. In the past couple of decades, everyone has become keenly aware of this, regardless of their satisfaction or otherwise with the modern synthesis. "You always feel like you're trying to cover your rear," says Love. "If you criticize, it's like handing ammunition to these folks." So don't criticize in a grandstanding way, says Coyne: "People shouldn't suppress their differences to placate creationists, but to suggest that neo-Darwinism has reached some kind of crisis point plays into creationists' hands," he says.[114]

This article therefore candidly admits a motive for suppressing criticisms of evolutionary theory: they have a "political concern" that they want to avoid "handing ammunition" to those they call the "fundamentalists" and "creationists."

Cannot Criticize Darwin

FINALLY, SINCE WE'VE BEEN ON THE TOPIC OF CHINA, HERE'S ONE FINAL comment to think about, from the Chinese paleontologist J. Y. Chen: "In China we can criticize Darwin, but not the government. In America, you can criticize the government, but not Darwin."[115]

These are not proponents of intelligent design. They are atheists and/or mainstream evolutionary scientists/scholars, telling us that scientists don't have the full freedom to express views that dissent from the standard evolutionary viewpoint.

But what would Neil deGrasse Tyson know about questioning Darwinism? Ask scientists and scholars who *have* tried to question Darwinian thinking, and you'll find out just how much freedom of expression there really is.

---

114. John Whitfield, "Biological theory: Postmodern evolution?" *Nature* 455 (17 September 2008), 281–284. See also: http://www.evolutionnews.org/2008/03/at_scoop_freelance_reporter_su004955.html.

115. J. Y. Chen quoted in Stephen C. Meyer, *Darwin's Doubt: The Explosive Origin of Animal Life and the Case for Intelligent Design* (New York: HarperOne, 2013), 52.

Episode 6:
Deeper, Deeper, Deeper Still

16.

Darwin's Flowers

Science as the New Sacred, and Failed Darwinian Predictions About Insects and Flowers

Casey Luskin

Cosmos Episode 6 is ostensibly about the miniature reality of the atom. Once again, there are some spectacular animations explaining the nature of atoms, and how they interact to form chemical bonds. We are also treated to animations of semi-steampunk-style machines, representing molecular machines in biology. Host Neil deGrasse Tyson has yet to dare tackle the question of how molecular machines might have evolved, but in this episode he does throw in a little tidbit supposedly highlighting a successful prediction of Darwinian evolution. He observes that "plants covered the surface of the Earth for hundreds of millions of years, before they put out their first flower," and then states that this led Darwin to make a prediction:

> On this basis of his theory of evolution through natural selection, Darwin speculated that somewhere on the island of Madagascar there must live flying insects with extraordinarily lengthy tongues—ones long enough to reach the pollen. No one had ever seen such a beast there. But Darwin insisted that an animal fitting this description must exist. It wasn't until more than 50 years later that Darwin was proven right.

And of course Tyson then notes that a particular hawk moth species exists on Madagascar that slurps pollen with its long tongue, "exactly as Darwin expected it would." He says, "There can be no stronger test of an idea than its predictive power," as if this fulfilled prediction regarding insects and flowers shows the amazingly successful predictive power of Darwinian evolution. However, a little critical thinking and investigation shows that nothing could be further from the case.

First of all, you wouldn't need to know anything about evolution to make the prediction Darwin supposedly made. Given the shape of the flower, if there weren't some insect capable of sipping its nectar, the flower could not attract insects capable of pollinating that species, and the plant would die out. There isn't some profound evolutionary principle at work here—a middle-school level knowledge of angiosperm reproduction could probably lead you to expect some insect must exist that enjoys this species's flowers and fosters its pollination.

Second, it actually turns out that Darwinian evolution has had great difficulties making good predictions when it comes to flowering plants (angiosperms) and insects. For example, Darwin generally predicted that species would appear gradually in the fossil record. However, it has been known for over 150 years that angiosperm groups appear abruptly, without clear evolutionary precursors, in the early Cretaceous period. Darwin himself called the "rapid development" of "higher plants" an "abominable mystery." This problem has persisted to the present day, as a paper in *Trends in Ecology and Evolution* states:

> research and analyses of different sources of data (e.g., fossil record and phylogenetic analyses using molecular and morphological characters), the origin of the angiosperms remains unclear. Angiosperms appear rather suddenly in the fossil record... with no obvious ancestors for a period of 80–90 million years before their appearance.[116]

One paper notes: "Darwin was deeply bothered by what he perceived to be an abrupt origin and highly accelerated rate of diversification of flowering plants in the mid-Cretaceous" and states, "Darwin's abominable mystery is about his abhorrence that evolution could be both rapid and potentially even saltational."[117]

---

116. Stefanie De Bodt, Steven Maere, and Yves Van de Peer, "Genome duplication and the origin of angiosperms," *Trends in Ecology and Evolution* 20 (2005), 591–597.

117. William E. Friedman, "The Meaning of Darwin's Abominable Mystery," *American Journal of Botany* 96 (2009), 5–21. This quote is from the abstract at: http://www.amjbot.org/content/96/1/5.short. In biology, saltation (from Latin, *saltus*, "leap") is a sudden

Somehow this abrupt appearance of flowering plants is never mentioned in *Cosmos* as a failed prediction of Darwinian theory. But Darwin made other bad predictions with regard to the diversification of flowering plants and insects—something much closer to the topic of this week's episode of *Cosmos*—which in fact derive directly from his attempts to explain away the abrupt appearance of flowering plants.

Writing in the *American Journal of Botany*, evolutionary botanist William Friedman observes that Darwin corresponded with the French paleontologist Gaston de Saporta who proposed that insects and flowering plants coevolved—i.e., they drove one another's evolution—and that Darwin hoped Saporta's ideas would explain the problematic abrupt appearance of flowering in the fossil record. As Friedman writes:

> ... Saporta (1873) was the first to suggest a critical and interdependent role of insects in the emergence and diversification of angiosperms.
>
> In his correspondence with Charles Darwin, Saporta elaborated on the theme of coevolutionary interdependence between insects and flowering plants—and this time, he tied his reading of the fossil record to issues associated with *rates of diversification*, the very essence of Darwin's "abominable mystery." In a letter that is notable for its brilliance and insights, Saporta explicitly proposed to Darwin that the rapid diversification of angiosperms was, in essence, a coevolutionary story tied to the origin of many major groups of insects.[118]

Don't miss Friedman's emphasis on "rates of diversification." The question here isn't merely about when these groups might have first appeared in the fossil record—after all, molecular clocks, trace fossils, and hard-body fossils give differing estimates of the first appearance of angiosperms and various insect groups.[119] Rather, we're looking at a much more concrete question: *When did flowering plants and insects experience their mass diversifications*—i.e., is there evidence for what Saporta called a "simultaneous" mass coevolution of insects and flowering plants?

---

change from one generation to the next, that is large, or very large, in comparison with the usual variation of an organism.

118. William E. Friedman, "The Meaning of Darwin's Abominable Mystery," *American Journal of Botany* 96 (2009), 15. Emphasis in original and internal citations omitted.

119. Casey Luskin, "Clocks versus Rocks," *Evolution News & Views* (17 January 2014), http://www.evolutionnews.org/2014/01/clocks_versus_r081291.html.

Darwin certainly liked Saporta's hypotheses. Friedman quotes Darwin replying to Saporta, calling the idea "splendid," and notes that "Darwin recognized the seminal importance of Saporta's hypothesis and was thus provided with a plausible, indeed powerful, mechanism to explain the rapid pace of early (meaning mid-Cretaceous, as of 1877) angiosperm diversification." Thus, Friedman writes:

> Darwin was at least partially responsible for stimulating the publication of the single most accepted hypothesis as to the cause of the rapid radiation of flowering plants, namely, their coevolution with insects. Moreover, Saporta's solution to Darwin's skepticism about the "sudden development" of flowering plants was the only explanation Darwin ever embraced to potentially explain a rapid diversification of early angiosperms as real.[120]

But did Saporta's prediction of the "simultaneous" coevolution of insects and planets stand the test of the evidence? A 1993 study in *Science* tried to systematically answer this question by looking at the fossil record. Authored by paleontologists Conrad Labandeira and J. John Sepkoski, Jr., the paper found that the great diversification of insects was apparently decoupled from the great radiation of flowering plants. As Labandeira and Sepkoski's paper in *Science* reported:

> The great radiation of modern insects began 245 million years ago and was not accelerated by the expansion of angiosperms during the Cretaceous period. The basic trophic machinery of insects was in place nearly 100 million years before angiosperms appeared in the fossil record.[121]

Citing a long list of evolutionary theorists, the paper observes that "The extraordinary diversity of living insects has been attributed by some workers to the diversity of angiospermous plants, which first appear as fossils in the Lower Cretaceous." Labandeira and Sepkoski note that while the exact timing of the appearance of the very first angiosperms is unclear, with estimates ranging back to the Triassic (indeed, in 2013 a potential flower-like pollen was reported from the Triassic),[122] the key question here

---

120. Friedman, "Darwin's Abdominal Mystery," 17. Internal citations omitted.

121. Conrad C. Labandeira and J. John Sepkoski, Jr., "Insect Diversity in the Fossil Record," *Science* 261 (July 16, 1993), 310.

122. "New fossils push the origin of flowering plants back by 100 million years to the early Triassic," *Science Daily* (1 October 2013), http://www.sciencedaily.com/releases/2013/10/131001191811.htm.

isn't the specific timing of the first potential appearance of angiosperms (the Triassic flower-like pollen, for example, is quite controversial),[123] but rather whether there exists a correlation between the mass diversification of flowering plants and the mass diversification of insects in the fossil record. Labandeira and Sepkoski write that if a record of such coevolution is to exist, "[i]t must be demonstrated that rates of diversification of insects increased as angiosperms diversified to dominate virtually all terrestrial plant communities."[124]

But this is not what they found. After providing a figure that tracks insect familial diversity over time, noting when angiosperms first appear and when they reach their "ascendancy," they note:

> *The more startling interpretation that can be drawn from the data (Fig. 4) is that the appearance and expansion of angiosperms had no influence on insect familial diversification....* the fossil data indicate that angiosperms experienced a tremendous radiation in all geographic regions during the Albian and Cenomanian stages of the middle Cretaceous. However, there is no signature of this event in the family-level record of insects. *Instead, the data in Fig. 4 suggest that insect diversification actually slackened as angiosperms radiated.*[125]

They conclude that "the post-Paleozoic radiation of insect families commenced more than 100 million years before angiosperms appeared in the fossil record" and this is true even for insects involved in pollinating flowering plants:

> Within the Insecta, orders that have radiated strongly during the Mesozoic and Cenozoic, such as the Coleoptera (beetles), Diptera (true flies), and Hymenoptera (wasps, ants, and bees), all apparently began their expansions during the Triassic and Jurassic (Fig. 2), long before the ascendancy of angiosperms.[126]

In other words, at the crucial level of families within insects, there is no evidence of "simultaneous" coevolution between flowering plants and

123. "Fossil Pollen Place Angiosperms in Triassic?," *Paleobotany and Plant Evolution* (6 October 2013), http://paleoplant.blogspot.de/2013/10/fossil-pollen-place-angiosperms-in.html.

124. Labandeira and Sepkoski, "Insect Diversity," 312.

125. Ibid, 313. Emphasis added.

126. Ibid, 313.

insects. The mass diversifications of these two groups are off by some 100 million years.

Reporting on Labandeira and Sepkoski's paper, the *New York Times* noted how it overturned conventional wisdom about the coevolution of insects and angiosperms:

> This assumption, elevated to conventional wisdom and taught in biology courses, has been challenged in a new, comprehensive study of insect fossils. The greatest expansion and diversification of insects, it has now been discovered, actually began 120 million years before the advent of angiosperms. If anything, when flowering plants proliferated, insect diversification slackened.
>
> [...]
>
> "The results contradict several notions about what macroevolutionary patterns can be seen among fossil insects and about how modern insect diversity can be interpreted," the scientists concluded in their report.
>
> [...]
>
> Dr. Leo J. Hickey, a paleobotanist at Yale University, said: "The results call into serious question some of our conceptions and preconceptions. All of us were quite comfortable with the idea that flowering plants must have had a major effect on insect diversity."[127]

So here we have a famous example where Darwin, and generations of evolutionary biologists after him, predicted on the basis of Darwinian theory that they'd find a tight correlation between the diversification of insects and flowering plants in the fossil record, as well as a fossil record showing the evolution of angiosperms. Instead, the predicted evidence for "coevolution" between angiosperms and insects has been elusive, and Darwin's "mystery" remains "abominable."

Where Is the Gratuitous Religion Bashing?

OF COURSE, NO EPISODE OF *COSMOS* WOULD BE COMPLETE WITHOUT Tyson's customary bashing of religious belief in the supernatural. In this episode, Tyson attributes the rise of science to

---

127. John Noble Wilford, "Long Before Flowering Plants, Insects Evolved Ways to Use Them," *New York Times* (3 August 1993), http://www.nytimes.com/1993/08/03/science/long-before-flowering-plants-insects-evolved-ways-to-use-them.html.

the idea that natural events were neither punishment nor reward from capricious gods. The workings of nature could be explained without invoking the supernatural.

Tyson tells us this non-theistic idea that the universe could be explained simply by the workings of natural laws came from Thales of Miletus. According to Tyson, the very idea of a "cosmos out of chaos, a universe governed by the order of natural laws, that we could actually figure out" was the "epic adventure" that Thales initiated.

As we've seen over and over again (see Chapters 6, 10, and 15), *Cosmos* wrongly depicts various early heroes of science as materialists when in fact they were theists who believed in a supreme God who created everything. Thales appears to be another example of this. According to Cicero, Thales argued that "water is the principle of all things; and that God is that Mind which shaped and created all things from water."[128]

Tyson next praises another ancient Greek philosopher, Democritus, as a "genius" who helped point us towards modern science. A member of Democritus' audience is shown asking, "You mean, that's it? That's all there is? Just a bunch of atoms in a void?" Democritus affirms that this is his belief. In Democritus, Tyson seems to have *finally* stumbled upon an early scientific thinker who arguably truly was a bona fide materialist and atheist. So where is *Cosmos* now in its efforts to enlist early scientific thinkers to bash religion—maybe 1 for 10 when it comes to citing people who actually were atheists? In any case, we must give credit where credit is due: Democritus was an atheist who rejected the pantheon of Greek gods. But science would not really get started until hundreds of years later when Christian scholars came along—who also rejected the capricious Greek gods in favor of a single God with a supreme mind that created an intelligible universe.

Spiritual Materialism

AT ONE POINT IN EPISODE 6, TYSON ENTERS A CATHEDRAL, AND FOR what feels like many seconds, an atom with orbiting electrons is portrayed overlapping with the stained glass windows of a cathedral. Why the

---

128. "Thales," Wikipedia, http://en.wikipedia.org/wiki/Thales.

strange emphasis on this imagery? If you follow New Atheist thinking and literature, the answer is simple. An important component of such thinking is the realization that science lacks the spiritual inspiration that religion provides for people. New Atheists are desperate to find ways to make science into a new form of human spirituality to replace religion. Thus, in this episode, Tyson gives us language like:

+ "Neutrinos from creation are within you."

+ "Go deeper into the wonder."

+ To explore this we'll need both "science" and "imagination."

+ "Atomic reincarnation."

If you read the literature of New Atheists, these subtle elements of the show instantly make sense. Carolyn Porco, a senior research scientist at the Space Science Institute in Boulder, Colorado has said, "If anyone has a replacement for God, then scientists do." Writing in the essay collection, *What's Your Dangerous Idea?*, she longs for the day that "science and formal religion will come to an end, when the role played by science in the lives of all people is the same as that played by religion today." Porco envisions "Einstein's Witnesses going door-to-door or TV evangelists passionately declaring the beauty of evolution." She concludes by stating that the new "sacred" shrines will be "astronomical observatories, the particle accelerators, the university research installations, and other laboratories where the high priests of science—the biologists, the physicists, the astronomers, the chemists—engage in the noble pursuit of uncovering the workings of nature." Porco hopes that "today's museums, exposition halls, and planetaria may become tomorrow's houses of worship," thereby replacing "formal religion." She ends by predicting, "'Hallelujah!' they will sing. 'May the force be with you!'"[129]

And that's exactly what we see in Episode 6: Today's sacred cathedrals become monuments not to God who creates atoms that build the cathe-

---

129. Carolyn Porco, "The Greatest Story Ever Told," *What's Your Dangerous Idea?: Today's Leading Thinkers on the Unthinkable*, John Brockman, ed. (New York: Harper Perennial, 2007), 153–155. See also: Casey Luskin, "False Spin," *Evolution News & Views* (17 March 2014), http://www.evolutionnews.org/2014/03/false_spin_ball083281.html.

dral, but to the atom. If that doesn't make you feel all warm and fuzzy inside, I suppose I don't know what will.

EPISODE 7:
THE CLEAN ROOM

17.

RELIGIOUS HISTORY,

DONE BADLY

Jay W. Richards

ADVANCE PUBLICITY FOR COSMOS RAN SO HOT THAT THE RATINGS were bound to disappoint.[130] After all, this was a science documentary stuck in the same time slot as *Masterpiece Classic*, *The Good Wife*, *Duck Dynasty*, and *River Monsters*.

Still, as an inveterate viewer of science documentaries, I hoped to see the very latest astronomical images and cutting-edge CGI used to explain arcane scientific concepts, and was not disappointed. When it was describing the inner workings of atoms and stars, or telling the story of how scientists came to decipher starlight, *Cosmos* was terrific.

Unfortunately, the series was sullied by the unscientific agendas of its producers.

This may have been inevitable. As previously noted, *Cosmos* is a reboot of the 1980 PBS series of the same name, hosted by late astronomer and atheist-evangelist Carl Sagan. The *Cosmos* redux is hosted by American astronomer and Sagan disciple Neil deGrasse Tyson, with help from Sagan's widow Ann Druyan and Seth MacFarlane, creator of the vulgar animated series *Family Guy* and *American Dad*. The atheist MacFarlane is one of Christianity's not-so-cultured despisers, so I expected a lot of bad materi-

---

130. "TV Ratings Sunday," *Zap2it* (17 March 2014), http://tvbythenumbers.zap2it.
   com/2014/03/17/tv-ratings-sunday-resurrection-the-mentalist-revenge-down-believe-
   tumbles-mediocre-premiere-for-crisis/245084/.

alist philosophy and shots at Christianity and religion.[131] What I did not expect was the mistreatment of history.

The historical interludes are always presented in flat cartoon animation, reflecting MacFarlane's role in the production. The format is fitting, since the treatment of history is so often cartoonish.

We've already discussed how the producers spent one fourth of the first episode telling a misleading story about Giordano Bruno (see Chapter 5), a sixteenth-century Dominican burned at the stake for a laundry list of unrepented heresies. He wasn't a scientist and had virtually nothing to do with the history of science. But *Cosmos* needed a martyr for science, and since there were none available, Bruno would have to do. The tale of Bruno cost the episode valuable airtime, however, so it had to give short shrift to trivial figures such as Copernicus and Galileo, who inconveniently died in their beds, unmartyred.

Cosmos's discussion of Bruno was so bad that even those who might be expected[132] to give the series rave reviews had the good sense to object. If the producers had simply checked the Wikipedia entry[133] for Giordano Bruno, they could have been spared the embarrassment.

In later episodes, Tyson claimed that the deeply religious Isaac Newton replaced God with gravity (see Chapter 10), and presented the ancient Chinese philosopher Mo Tze (Mozi) as an early skeptic who was apparently "against faith" (see Chapter 15). (His religious views were actually much less cartoonish.)[134] And so it has gone, with a rotten historical Easter egg in many of the episodes.

I was eager to find out what Sagan's widow, MacFarlane, and Tyson would do in Episode 7, which aired on April 20, Easter Sunday. Maybe they'd take a brief excursus to explain that science has determined that

131. Stacey G. Woods, "Hungover with Seth MacFarlane," *Esquire* (18 August 2009), http://www.esquire.com/features/the-screen/seth-macfarlane-interview-0909.

132. Josh Rosenau, "Why Did Cosmos Focus on Giordano Bruno?" *Science League of America* (18 March 2014), http://ncse.com/blog/2014/03/why-did-cosmos-focus-giordano-bruno-0015457.

133. "Giordano Bruno," Wikipedia, http://en.wikipedia.org/wiki/Giordano_Bruno.

134. "Mohism," *Stanford Encyclopedia of Philosophy* (21 October 2002, revised 14 May 2014), http://plato.stanford.edu/entries/mohism/#religion.

the resurrection of Jesus was a myth. Or maybe they would serve up the old chestnut about Easter being a lightly christened pagan fertility festival. That one is also not true.[135]

But they were far craftier. In the Easter Sunday episode about how modern scientists determined the true age of the Earth (and about the tenuously related twentieth-century controversy over leaded gas), they inserted a segment about Christmas. We learn that the holiday celebrated by a couple billion Christians is really a camouflaged take-over of Saturnalia, the High Holy Day when ancient Romans celebrated Saturn, the god of agriculture. How is this relevant? Well, Saturn is also the name of a planet, which is part of the solar system, which is part of the cosmos.

Like several of *Cosmos*'s previous detours into history, this one also leads off a cliff. Perhaps for the writers, the notion that Christmas is really a purloined pagan festival is one of those claims that is too good not to be true. And too good to need verification.

As the story goes, there was a Roman celebration of the end of the autumn growing season—corresponding to the winter solstice—when people treated each other kindly, helped the poor, and so forth.[136] Various church fathers and Christian figures needed good branding for their new religion, and decided to gussy up Saturnalia with Christian garb. Perhaps they thought that the drunk Saturnalians would eventually forget what all the commotion was about and find themselves raising a glass to the birth of Jesus.

As it happens, there was another Roman celebration that competed for the same slot on the calendar, *Sol Invictus*, the birthday of the unconquered Sun. By invoking Jupiter and a Sun King, later Roman emperors hoped to use this holiday to further their claim to divinity.

In the eighteenth and nineteenth centuries, *Sol Invictus* was sometimes identified as the ancestor of Christmas. But this is now widely disputed

135. Andrew McRoy, "Was Easter Borrowed from a Pagan Holiday?," *Christian History* (2 April 2009), http://www.christianitytoday.com/ch/bytopic/holidays/easterborrowedholiday.html.

136. Matt Salusbury, "Did the Romans Invent Christmas?" *HistoryToday* 59 no. 12 (December 2009), http://www.historytoday.com/matt-salusbury/did-romans-invent-christmas.

by historians—again Wikipedia has the story.[137] In fact, there's some evidence that it may have been an attempt to create a Roman, and pagan, alternative to Christmas.[138]

Most likely, the dating of Christmas has a prosaic theological explanation. It's when early Christian thinkers thought Jesus was born. For reasons that aren't important here, they inferred that Jesus' conception took place on March 25. As Matt Salusbury explains in *History Today*, "Early ecclesiastical number-crunchers extrapolated that the nine months of Mary's pregnancy following the Annunciation on March 25 would produce a December 25 date for the birth of Christ."[139]

Of course, this leads us quite far afield from the obvious question: Why on earth did the producers of *Cosmos* decide to repeat this doubtful story about the origin of Christmas? My guess is that Christianity is one of their targets and for beating some dogs any old stick will do.

137. "Sol Invictus," Wikipedia, http://en.wikipedia.org/wiki/Sol_Invictus.

138. William Tighe, "Calculating Christmas," *Touchstone* (December 2003), http://touchstonemag.com/archives/article.php?id=16-10-012-v.

139. Matt Salusbury, "Did the Romans Invent Christmas?" *History Today* 59 no. 12 (December 2009), http://www.historytoday.com/matt-salusbury/did-romans-invent-christmas.

Episode 8:
Sisters of the Sun

18.

A Speck of Sand

So We're Made of "Stardust"—But Is That All?

Casey Luskin

Episode 8 of *Cosmos* is highlighted by a fascinating discussion of how scientists have determined the compositions of stars, and what happens when different types of stars die, producing phenomena like supernovae, black holes, or even more rarely, a hypernova. The science-lover and sci-fi fan in me was glued to the screen. After recounting how the pioneering astronomer Cecilia Payne-Gaposchkin bravely bucked the consensus of the male-dominated astrophysical community in the 1920s, showing that the sun was made primarily of hydrogen and helium, Tyson remarked: "The words of the powerful may prevail in other spheres of human experience, but in science, the only thing that counts is the evidence, and the logic of the argument itself." If only that were always true!

As we saw in response to Episode 6 (see Chapter 16), plenty of scientists have said that when it comes to neo-Darwinism, the presentation of counter-evidence often isn't tolerated. The history of science shows this in other fields as well. Some 15 years before Payne-Gaposchkin published her discoveries, Alfred Wegener was widely ridiculed and roundly rejected because of his belief in continental drift.[140]

---

140. Casey Luskin, "For Intelligent-Design Advocates, Lessons from the Debate over Continental Drift," *Evolution News & Views* (14 June 2012), http://www.evolutionnews.org/2012/06/for_intelligent060911.html.

There was some decent evidence for continental drift even in Wegener's time—e.g., fit of the continents, fossils on matching places across the Atlantic—both of which are still regularly cited in geology textbooks as evidence for continental drift. Anti-drift scientists had weak rebuttals to the evidence Wegener cited, appealing to the migration of organisms across oceans or over ancient land bridges, and claiming the correlations were basically mere coincidence. The case for continental drift wasn't nearly as strong as it was after paleomagnetic data was discovered decades later, but that doesn't mean there was zero good evidence in Wegener's time, and it doesn't mean Wegener deserved the nasty dismissal or the ridicule he received. It also doesn't mean scientists thought objectively about the situation—that's an important point because Tyson in *Cosmos* promotes a naïve view of science that scientists are like perfectly objective robots.

In any event, the great historian of science Thomas Kuhn said, "No part of the aim of normal science is to call forth new sorts of phenomena; indeed those that will not fit the box are often not seen at all. Nor do scientists normally aim to invent new theories, and they are often intolerant of those invented by others."[141]

Nonetheless, most of Episode 8 is great science, history, and science communication. It wasn't until the very end that Tyson couldn't help but get all fuzzy and materialistic. He enthuses: "Our ancestors worshiped the sun. They were far from foolish. It makes good sense to revere the sun and stars because we are their children."

He goes on to say that all of our elements, our society, everything about us is "stardust." He explains, "We are made by the atoms and the stars" and "our matter and our form are forged by the great and ancient cosmos of which we are a part." But is that the end of the story, as Tyson makes it sound? Is our "form" forged by the cosmos alone?

We've already dealt with Tyson's failure to recognize that producing a life-friendly planet like our own—whether from "stardust" or something else—requires an incredible amount of fine-tuning.[142] And that fine-tuning

---

141. Thomas S. Kuhn, *The Structure of Scientific Revolutions*, 3rd ed. (Chicago: University of Chicago Press, 1996), 24.
142. See Chapter 13.

by itself won't even get you life. I'm sorry, but supernova explosions don't produce—in any way, shape or form—the conditions necessary for generating the complex and specified language-based code that underlies all life on Earth.

Tyson takes no notice of the reality that we live on a privileged planet,[143] and instead opts throughout *Cosmos* to subtly promote a position ultimately more like that of popular "Science Guy" Bill Nye, who said:

> I'm insignificant.... I am just another speck of sand. And the Earth really in the cosmic scheme of things is another speck. And the sun an unremarkable star. Nothing special about the sun. The sun is another speck. And the galaxy is a speck. I'm a speck on a speck orbiting a speck among other specks among still other specks in the middle of specklessness. I suck.[144]

But this simply ignores the mountains of evidence for the fine-tuning[145] needed for our cosmos, galaxy, solar system, and planet to allow a life-friendly habitat. So no, we are not "forged" by the cosmos. And while the life-friendly cosmos points to design, cosmic fine-tuning alone is necessary but not sufficient to generate life. If there really were nothing more than Tyson's materialistic universe, we wouldn't be here to talk about it.

143. *The Privileged Planet*, http://www.theprivilegedplanet.com/.

144. "Bill Nye Speaks at the 2010 AHA Conference," YouTube video (5 June 2010), http://www.youtube.com/watch?v=S4dZWbFs8T0.

145. Casey Luskin, "Roger Penrose on Cosmic Fine-Tuning," *Evolution News & Views* (12 April 2010), http://www.evolutionnews.org/2010/04/roger_penrose_on_cosmic_finetu033691.html.

EPISODE 9:
THE LOST WORLDS OF
PLANET EARTH

19.

Catastrophes!!!

Marred by the Now Familiar Rigid Ideology,
Cosmos Tackles Geology and Climate

Jay W. Richards

THE NINTH INSTALLMENT OF COSMOS, "THE LOST WORLDS OF PLANet Earth," focuses mostly on geology. The episode is a representative sample of the series. It mixes one part illuminating discussion of scientific discoveries, one part fanciful, highly speculative narrative, and one part rigid ideology disguised as the assured results of scientific research. This would be all right if the different ingredients were clearly distinguishable, like a marble rye. But they are mixed so thoroughly that the casual viewer can have little idea where the evidence leaves off and the speculation and ideology begin.

First, the scientific discovery: This episode tells the inspiring stories of Alfred Wegener and Marie Tharp, the geologists whose work helped give rise to our understanding of large scale geology and plate tectonics. Wegener was one of the great scientific dissenters of the twentieth century. In fact, he died while he was still considered a "denier" of the established scientific consensus. But his notion of continental drift eventually won the day, once the evidence and the mechanisms for plate tectonics became hard to deny.

Host Neil deGrasse Tyson admits that the case of Wegener (and to a lesser extent, Tharp) shows that scientists can be as guilty of prejudice and group think as a gaggle of middle schoolers with low self-esteem. (He didn't use that imagery, but you get the point.) Unfortunately, such a frank admission of group think among scientists is never applied to the current intellectual orthodoxies that form the backbone of the *Cosmos* series.

The speculative scenarios in this episode principally involve claims about how extinction events helped evolution along. There's nothing wrong with speculation as long as it is identified as such. But in "The Lost Worlds of Planet Earth," we receive no such viewer advisories. The contested hypothetical details about the Permian-Triassic Extinction, for instance, which are described with appropriate tentativeness on Wikipedia, get flattened into a narrative that seems as certain and uncontroversial as a basic time line of World War II.[146]

In this episode, the hard-edged ideology comes mostly in the form of extreme environmental catastrophism. We learn of the negative effects of methane, atmospheric carbon dioxide, and global warming in the ancient past. No conscious viewer could fail to expect that a sermon about climate change is not far behind.

But the sermon unwittingly contradicts the general thrust of the episode. The CGI and the narration make it clear that the present era (since the last Ice Age) is far more peaceful and life-friendly than previous Ice Ages and the super-volcanic infernos of the distant past. For instance, Tyson tells of a super volcano in Siberia that shut down "the circulatory system of the ocean," and describes periods in which the global climate fluctuated between extreme heat and extreme cold. If anything is obvious, it is that *man had nothing to do with these events in the distant geologic past, and nothing that extreme is happening now.*

Nevertheless, this serves as a set-up to hector viewers who continue to dredge up the carbon planted in the Earth's crust, rather than using the abundant and "free" sunlight all around us.

Now it's reasonable to think that the human contributions of carbon dioxide to the atmosphere (about two parts per million per year) in the last couple of centuries is having some effect on the climate. And there

---

146. "Permian–Triassic extinction event," Wikipedia, http://en.wikipedia.org/wiki/
Permian%E2%80%93Triassic_extinction_event.

are plenty of reasonable people who think that those effects are more bad than good, even though the main presumed effect—global warming—is not presently happening. (See Chapter 25.) But it is simply bizarre to compare the conditions that led to massive extinctions in Earth's distant past with our current climate quandaries. And to contrast the "costly" energy from carbon sources (coal, oil, natural gas) with "free" energy from the sun displays a staggering level of economic illiteracy.

If we had the ability to convert sunlight to usable energy at no cost— that's what "free" means—then no one would be burning coal or refining oil for energy. If this is not clear to you, stop and think about it for a minute. If it cost ten cents to extract a barrel of oil from the ground and refine it, that would still be far more expensive than usable solar energy—which remember, would be free. So no one would bother to extract oil for energy.

The reason we still use oil, natural gas, and coal is that in the real world, it is by far the most economical choice for many uses. Converting sunlight to usable energy requires expensive technology and doesn't provide much energy. That doesn't mean that fossil fuels will never be displaced by other, "renewable" sources of energy. It's just that, at the moment, such sources are not economically feasible for most widespread uses.[147]

It is a dangerous ideology that resists even such rudimentary common sense.

---

147. Robert Bryce, "Don't Count Oil Out," *Slate* (14 October 2011), http://www.slate. com/articles/technology/future_tense/2011/10/oil_and_gas_won_t_be_replaced_by_ alternative_energies_anytime_so.html.

20.

CONTINENTAL DRIFT

FOR NEIL TYSON AND *COSMOS*, SERIOUS SCIENTIFIC CONTROVERSIES ARE ALL A THING OF THE PAST

Casey Luskin

T HE NINTH INSTALLMENT OF *COSMOS* COVERS ONE OF MY FAVORITE topics, plate tectonics, which is also one of the best supported theories in geology. The episode explains how crust is created at mid-ocean ridges, and then gets subducted back into the mantle. The animations and other visual explanations are great as usual.

Tyson, however, couldn't resist throwing in his usual dose of theology. Earthquakes, he explained, are caused by tectonic forces "not because somebody misbehaved and is being punished. It's due to random forces that are governed by the laws of nature." The interjection seemed totally out of place if all you were trying to do is teach the public about science, as opposed to promoting a materialist worldview. But as we've seen, teaching materialism is indeed a large part of Tyson's agenda.

In discussing plate tectonics, *Cosmos* offers a nice retelling of the stories of Abraham Ortelius, and later Alfred Wegener and Marie Tharp, who proposed early ideas about continental drift. Wegener, the famous German meteorologist, was harshly ridiculed for advocating such a theory.[148] Neil Tyson explains that Wegener cited as evidence not only the way

---

148. Casey Luskin, "For Intelligent-Design Advocates, Lessons from the Debate over Continental Drift," *Evolution News & Views* (14 June 2012), http://www.evolutionnews. org/2012/06/for_intelligent060911.html.

continents look as if they could fit together like a jigsaw puzzle, but also matching fossils on both sides of the Atlantic. I discussed some of this same evidence in my response to Episode 8 (see Chapter 18).

Tyson observes, "Most geologists ridiculed Wegner's hypothesis of continental drift" and preferred "imaginary land bridges to explain away Wegener's evidence." But since Wegener could offer no mechanism to explain how the continents might have plowed through the ocean floor, his theory was not accepted. In Chapter 18, I explained what happened:

> Obviously the case for continental drift [in Wegener's day, the early 1900s] wasn't nearly as strong as it was after paleomagnetic data was discovered decades later, but that doesn't mean there was zero good evidence in Wegener's time, and it doesn't mean Wegener deserved the nasty dismissal or the ridicule he received. It also doesn't mean scientists behaved objectively about the situation—that's an important point because Tyson in *Cosmos* promotes a naïve view of science that scientists are like perfectly objective robots.

In Episode 9, it's almost as if Tyson heard my concerns—which of course is impossible since these programs were undoubtedly all produced long before the first one aired. Immediately after discussing Wegener, Tyson offered the following wonderful statement acknowledging the fallibility of scientists:

> Scientists are human. We have our blind spots and prejudices. Science is a mechanism designed to ferret them out. The problem is we aren't always faithful to the core values of science.

Exactly. It's heartening to hear Tyson acknowledge this point. In light of it, I would like to soften my prior statement about Tyson's "naïve view." Yet despite his concession, something still bothered me about *Cosmos*'s presentation of how science works. It wasn't until a few hours after Episode 9 ended that I put my finger on it.

Yes, Tyson should be commended for recognizing many past instances where the scientific consensus turned out to be wrong. But his discussions of past scientific errors are always just that—in the past.

Cosmos has done a wonderful job of recalling how old mistaken ideas were overturned—ideas about geocentrism, stellar composition, continental drift, whether lead is dangerous to human health, and more. *However, these are all tales from the annals of scientific history. Cosmos* presents current scientific thinking as if it were all correct, with everything figured out, never recognizing that today too there are serious questions about dominant scientific ideas. Certainly Tyson never discusses evidence that challenges the prevailing evolutionary view.

Permian Extinction

IN EPISODE 9, FOR EXAMPLE, HE SPENDS QUITE A BIT OF TIME DISCUSSING the Permian extinction, the largest such event in the history of life on Earth. He is right that many different kinds of animals died out at the end of the Permian. Why did it happen? Tyson tells a convoluted tale involving highly complex causes. He presents these without qualification, as if scientists have solved the mystery and all agree on the same story. Not true. In reality, the triggers of the Permian extinction are still very much an unsolved enigma. Tyson never acknowledges that its causes are very difficult even for modern scientists to decide. That is because of the mysterious grab-bag diversity of the organisms that perished, from deep sea to terrestrial.

Though this mass extinction very likely involved the acidification of the oceans and atmosphere, the precise reasons for it are still unclear to biologists. Proposals range from asteroid impacts, to mass earthquakes, tsunamis, and flood basalt volcanism, to a global freeze, to a runaway greenhouse situation, to methane being released from undersea gas hydrates. Many other hypotheticals are cited. They may all be possible, but if the entire event took place in a window of just 60,000 years, as seems to be the case, what are the odds of all these factors conspiring to happen around the same time?[149]

Try doing a Google search for "Permian extinction debate," and you'll see just how much controversy remains. A 2012 book, *Earth and Life*, explains:

---

149. Zoe Mintz, "Permian Mass Extinction Took 60,000 Years, Siberian Traps May Have Triggered 'Really Rapid' Wipeout," *International Business Times* (11 February 2014), http://www.ibtimes.com/permian-mass-extinction-took-60000-years-siberian-traps-may-have-triggered-really-rapid-wipeout.

The quasi-perennial debate on the causes of the end-Permian extinction is shifting from a simple uni-parameter to a multi-parameter cause in which short-term and long-term changes of each factor concurred to disrupt the global disequilibrium of Late Permian life. The debate has given rise to a thousand papers or more, often based on several datasets of differing nature (litho-, bio-, chrono-, chemo- and magnetostratigraphy), leading to contrasting or partly overlapping solutions for this enigma. Moreover, it is not unusual to find a set of data collected in one stratigraphic section contrasting with data of the same nature collected from the same section by different researchers. This puts geologists at risk of shifting towards a "nihilistic" scientific position.[150]

Likewise, a 2012 paper in *Annual Reviews of Earth and Planetary Science* says this:

> Few events in the history of life pose greater challenges or have prompted more varied speculation than the end-Permian mass extinction [~252 million years ago (Mya)], the greatest biodiversity crisis in the history of animal life....
>
> During the past decade, debate over the causes of the end-Permian mass extinction has focused on three geological triggers. These are (a) bolide impact; (b) overturn or upwelling of deep water in a stratified, anoxic ocean; and (c) flood basalt volcanism.[151]

Cosmos mentions many of these potential triggers, but gives no hint of how much disagreement there is. Clearly, the Permian extinction happened—the fossil record attests to that—and it's not unreasonable to think that some of the causes that have been offered could be correct. However when *Cosmos* confidently tells a story like this, without qualification, it paints an inaccurate picture of what scientists actually say, and whitewashes all the disagreement on the topic.

As a second example, this episode presents as solid fact the notion the ancestor of all mammals was a small weasel-like creature that lived in what would later be Newark, New Jersey. It's a cute story, but the history of

150. John A. Talent, *Earth and Life: Global Biodiversity, Extinction Intervals and Biogeographic* (Dordrecht, Springer, 2012), 721. Emphasis added.

151. Jonathan L. Payne and Matthew E. Clapham, "End-Permian Mass Extinction in the Oceans: An Ancient Analog for the Twenty-First Century?" *Annual Reviews of Earth and Planetary Science* 40 (2013), 89–111, http://www.annualreviews.org/doi/pdf/10.1146/annurev-earth-042711-105329.

mammals is a more complicated subject than *Cosmos* lets on. Tyson later notes that mammals diversified at the end of the Cretaceous after the dinosaurs died out. But there's no mention of the abruptness of this diversification, in what has been called by some a "mammal explosion."[152] As Niles Eldredge wrote in the 1980s: "There are all sorts of gaps: absence of gradationally intermediate 'transitional' forms between species, but also between larger groups—between, say, families of carnivores, or the orders of mammals."[153]

A 2000 paper in *Nature* describes another conundrum facing researchers who study the origin of mammals:

> [M]any orders of mammals and birds are now thought to have originated long before the end-Cretaceous extinction, which occurred 65 Myr ago and which was thought previously to have been the signal for their radiation. If the new timescale can be trusted, these findings present a puzzle and a warning. The puzzle is the absence of fossils. Why have we not found traces of these lineages in their first tens or even hundreds of millions of years?[154]

Now you might respond that *Cosmos* has limited airtime and can't get into every unresolved question. That's true, but nearly every episode does devote a lot of time to discussing scientific controversies. The problem is, it only discusses controversies that were resolved many decades (if not multiple centuries) ago, and never mentions current debates with the scientific community—especially when they pertain to the evolution of life.

Savannah Hypothesis

A FURTHER EXAMPLE: TYSON REPEATS THE OLD SAVANNA HYPOTHESIS of human origins. According to this idea, when Central America formed tectonically, it separated the Pacific and Atlantic Oceans, changing ocean currents, and ultimately creating a drier climate in Africa. As the story

---

152. Robert Folzenlogen, "The Mammal Explosion," *Nature's Blog* (6 July 2008), http://naturesblog.blogspot.com/2008/07/mammal-explosion.html.

153. Niles Eldredge, *The Monkey Business: A Scientist Looks at Creationism* (New York: Washington Square Press, 1982), 65.

154. Andy Purvis and Andy Hector, "Getting the measure of biodiversity," *Nature* 405 (11 May 2000), 212–19, http://www.nature.com/nature/journal/v405/n6783/full/405212a0.html.

goes, the drier climate in Africa diminished the woodlands and the forest, forcing our forebears out of their trees. According to Tyson, this is what drove human evolution. Showing an animated scene of an australopithocene-like creature standing in a dead tree amid a dry landscape, Tyson explains at length:

> But then it got colder, and the trees thinned out. Broad grasslands sprang up. And our ancestors were forced to traverse them in search of food. You needed a totally different skillset to make it on the savanna.
>
> In the old days you could sit perched on your tree branch and watch the big cats from a safe distance. Now you were playing on the same dangerous field.
>
> The survivors were those who evolved the ability to walk great distances on their hind legs and to run when necessary. This changed the way they looked at the world. Hands and arms were no longer tied up with walking. They were free to gather food, and pick up sticks and bones. These could be used as weapons and tools.

He summarizes:

> Think of it: A change in the topography of a small piece of land half a world away reroutes ocean currents. Africa grows colder and drier. Most of the trees can't withstand the new climate. The primates who lived in them have to seek other homes. And before you know it, they're using tools to remake the planet. The Earth has shaped the course of human destiny.

So the climate got drier in Africa, and "before you know it" human intelligence evolved. Is this compelling? Of course not. Tyson leaves entirely unmentioned any biological explanation of what had to change for the chimp-like brains of our supposed australopithecine ancestors to achieve human intelligence. That's a gargantuan omission. Nor does he mention that the savanna hypothesis has prominent detractors among evolutionary paleoanthropologists.

As my co-authors and I explained in the book *Science and Human Origins*, back in 2001 the species *Orrorin tugensis*[155] was being promoted[156]

---

155. Casey Luskin, "The Fragmented Record of Early Hominins," *Evolution News & Views* (31 July 2012), http://www.evolutionnews.org/2012/07/the_fragmented062681.html.

156. John Noble Wilford, "Fossils May be Earliest Human Link," *New York Times* (21 July 2001), http://www.nytimes.com/2001/07/12/world/fossils-may-be-earliest-human-link.html.

as possibly the "earliest human link." It was called potentially "the earliest known ancestor of the human family."[157] It was also thought that this creature might have walked upright. But because the fossil was deposited in a wooded environment, it threatened the savanna hypothesis, according to which upright walking evolved so our supposed ancestors could stand tall and see over the grass. Martin Pickford, one of the investigators who studied *Orrorin*, explained:

> The fact that *Orrorin* is found with other fauna that indicates a wooded to forested environment, tends to refute the "savanna" hypothesis of human origins. It seems more likely now that bipedalism originated from an arboreal ancestor, rather than via a knuckle-walking ground dweller similar to chimpanzees. Thus, previous palaeoecological and palaeoenvironmental scenarios of human origins will need to be reconsidered in light of the new data.[158]

Then, when "Ardi" was unveiled in 2009, she became the new human-ancestor du jour, and again the savanna hypothesis was ready to be discarded. Because the data suggest she walked upright in a wooded, forest environment, *Time* Magazine explained that the savanna hypothesis could no longer be considered correct:

> This tableau demolishes one aspect of what had been conventional evolutionary wisdom. Paleoanthropologists once thought that what got our ancestors walking on two legs in the first place was a change in climate that transformed African forest into savanna. In such an environment, goes the reasoning, upright-standing primates would have had the advantage over knuckle walkers because they could see over tall grasses to find food and avoid predators. The fact that Lucy's species sometimes lived in a more wooded environment began to undermine that theory. The fact that Ardi walked upright in a similar environment many hundreds of thousands of years earlier makes it clear that there must have been another reason.[159]

---

157. John Noble Wilford, "On the Trail of a Few More Ancestors," *New York Times* (8 April 2001), http://www.nytimes.com/2001/04/08/world/on-the-trail-of-a-few-more-ancestors.html.

158. Martin Pickford, "Fast Breaking Comments," *Essential Science Indicators* (December 2001), http://www.esi-topics.com/fbp/comments/december-01-Martin-Pickford.html.

159. Casey Luskin, "Artificially Reconstructed 'Ardi' Overturns Prevailing Evolutionary Hypotheses of Human Evolution," *Evolution News & Views* (2 October 2009), http://www.evolutionnews.org/2009/10/artificially_reconstructed_ard026201.html.

Writing in *Science*, prominent paleoanthropologists, including Tim White, observed that Ardi's depositional environment "is not easily accommodated by an environmentally deterministic view that involves globally shrinking forests spawning savanna-striding hominids. We contend that this narrative is now undermined by the totality of data from [the location Ardi was discovered]."[160]

Other technical papers have challenged the savanna hypothesis. A 2012 paper in *Philosophical Transactions of the Royal Society B* observes that the appearance of *Homo*—which marks the crucial point where big brains first appear—didn't even take place on the African savanna:

> The appearance of the genus *Homo*, and subsequently of *H. erectus* (or *ergaster*), was historically associated with the expansion of the savanna. However, recent reinterpretations of the palaeoclimate record have questioned this hypothesis. Recent re-evaluations of the African palaeoclimate data suggest that pulsed changes may be more important than long-term trends. Moreover, these analyses suggest that the periods associated with this step-change in encephalization may have occurred during a wet rather than a dry period.[161]

Another paper in *Palaeogeography, Palaeoclimatology, Palaeoecology* found that there have been four dry periods in Africa over the past four million years, meaning that if climate change did drive the evolution of human intelligence, then "any such link is complex," and not nearly as tidy as Tyson presents it.[162]

Another paleoclimate paper (ultimately published in *Quaternary Science Reviews*) states:

> We analyzed published records of terrigenous dust flux from marine sediments off subtropical West Africa, the eastern Mediterranean Sea,

160. Tim D. White, et al, "Response to Comment on the Paleoenvironment of Ardipithecus ramidus," *Science* 328 no. 5982 (28 May 2010), 1105, http://www.sciencemag.org/content/328/5982/1105.5.full.

161. Susanne Shultz, Emma Nelson, and Robin I. M. Dundar, "Hominin cognitive evolution: identifying patterns and processes in the fossil and archaeological record," *Philosophical Transactions of the Royal Society* 367 no. 1599 (25 June 2012), 2130–2140, http://rstb.royalsocietypublishing.org/content/367/1599/2130.full.

162. René Bobe and Anna K. Behrensmeyer, "The expansion of grassland ecosystems in Africa in relation to mammalian evolution and the origin of the genus *Homo*," *Palaeogeography, Palaeoclimatology, Palaeoecology* 207 no. 3–4 (20 May 2004), 399–420, http://www.sciencedirect.com/science/article/pii/S0031018204000495.

and the Arabian Sea, and lake records from East Africa using statistical methods to detect trends, rhythms and events in Plio-Pleistocene African climate. The critical reassessment of the environmental significance of dust flux and lake records removes the apparent inconsistencies between marine vs. terrestrial records of African climate variability. Based on these results, major steps in mammalian and hominin evolution occurred during episodes of a wetter, but highly variable climate largely controlled by orbitally-induced insolation changes in the low latitudes.[163]

Again, my concern isn't that *Cosmos* discusses the savanna hypothesis, but that it promotes this disputed theory uncritically, wrongly suggesting that scientists all agree on the causes of one of the most profound questions science can tackle—the origin of the human mind. *Cosmos* would do better to admit where ideas about human evolution in fact are lacking in evidence, or are controversial.[164]

Speaking of trees, Tyson might even mention that alternative hypotheses like intelligent design explain some of the very data that has puzzled evolutionary biologists. "Trees evolved a way to defy gravity. A plant molecule evolved that was both strong and flexible. A material that could support a lot of weight, but could bend in the wind," Tyson asserts in this episode. That material is lignin. Tyson explains that "lignin had a downside. It was hard to swallow." More precisely, lignin is notoriously difficult for organisms to break down and metabolize. If Darwinian processes are so adept at evolving new traits, why has this problem been so hard for evolution to solve? *Cosmos* doesn't mention this, but the hypothesis of pro-intelligent design scientists may hold the answer.

Matti Leisola (a specialist in enzymology and enzyme engineering, formerly at Aalto University School of Chemical Technology, Finland), Ossi Pastinen (also at Aalto University in the School of Chemical Technology), and Douglas Axe (Biologic Institute) authored a 2012 peer-reviewed paper in *BIO-Complexity*, "Lignin: Designed Randomness." They argue that the reason no organisms can digest lignin is to allow the buildup of humus

---

163. M. H. Trauth, J. C. Larasoaña and M. Mudelsee, "Trends, rhythms and events in Plio-Pleistocene African Climate," *Geophysical Research Abstracts* 11 (2009), http://meetingorganizer.copernicus.org/EGU2009/EGU2009-2362.pdf.

164. Casey Luskin, "A Big Bang Theory of *Homo*," *Evolution News & Views* (13 August 2012), http://www.evolutionnews.org/2012/08/a_big_bang_theo063141.html.

in the soil, which in turn permits plant growth, all resulting in life that depends on plants. Lignin forms the basis of an ecological fine-tuning argument, pointing to a designed ecosystem. But since *Cosmos* has thus far refused to mention fine-tuning, it's probably not interested in discussing this hypothesis.

In their paper, Leisola, Pastinen, and Axe recognize that lignin poses a conundrum for Darwinian theory, which tends to suppose that new molecular functions readily evolve. Instead, they find:

> The Darwinian account must somehow reconcile 400 million years of failure to evolve a relatively modest innovation—growth on lignin—with a long list of spectacular innovations thought to have evolved in a fraction of that time.[165]

They thus ask: "How can microorganisms have failed to exploit lignin as an energy source while much less evolvable species have, on innumerable occasions, acquired solutions to problems that appear to be considerably harder?" In their view:

> That tension vanishes completely when the design perspective is adopted. Terrestrial animal life is crucially dependent on terrestrial plant life, which is crucially dependent on soil, which is crucially dependent on the gradual photo- and biodegradation of lignin. Fungi accomplish the biodegradation, and the surprising fact that it costs them energy to do so keeps the process gradual. The peculiar properties of lignin therefore make perfect sense when seen as part of a coherent design for the entire ecosystem of our planet.[166]

They conclude that lignin makes an argument for design not just in microbiology, but also in ecology:

> Perhaps the oddest aspect of this is that Darwin's theory is unable to make sense of a situation that otherwise makes perfect sense. If life is the product of intelligent design, it stands to reason that the whole design must be considered—not just the functions of molecules and cells and tissues and organs and organisms, but also the functions of entire ecosystems, all the way up to the global ecosystem.[167]

---

165. Matti Lessola, Ossi Pastinen and Douglas D. Axe, "Lignin—Designed Randomness," *BIO-Complexity* 2012 (2012), http://bio-complexity.org/ojs/index.php/main/article/view/BIO-C.2012.3.

166. Ibid.

167. Ibid.

Whatever you believe about evolution and intelligent design, the origin of humans, or the future of climate change, dissent deserves to be heard. It's a shame that *Cosmos* presents current science as a monolithic enterprise, where the current "consensus" on basic ideas is fully correct. Such a view scuttles debate. Indeed, debate is hardly acknowledged to occur at all anymore.

Episode 10:
The Electric Boy

21.

THE FAITH OF FARADAY

& MAXWELL

THE WHITEWASH CONTINUES: CONCEALING
THE SOURCES OF SCIENTIFIC INSPIRATION FOR
MICHAEL FARADAY, JAMES CLERK MAXWELL

David Klinghoffer

EPISODE 10 OF COSMOS, "THE ELECTRIC BOY," IS BEAUTIFULLY DONE, and quite moving. I got choked up more than once, especially by its telling of how the self-educated genius Michael Faraday suddenly fell victim to memory loss and depression at age 49. Faraday persevered admirably despite it. At last he saw his ideas vindicated, recast in mathematical terms, by a younger man, James Clerk Maxwell.

Cosmos is unstinting in its praise for Faraday, without whose elucidation of electromagnetism and other discoveries we might still be living today, says Neil deGrasse Tyson, as our forebears did in the nineteenth century. Actually, what Tyson says is that without him we could be living as they did in the *seventeenth century*—but given that Faraday was born in 1791 and died 1867, I don't see how his never having lived would turn the clock back *that* far.

Do I need to tell you what Tyson leaves out of this uplifting story? Yes, of course, it's any admission of how Faraday and Maxwell were *inspired in their science by their Christian commitments*. In fact, you could hardly ask for two names of great scientists whose work was more influenced by passionate religious views, including the vision that nature reflects a single unified cosmic design.

Faraday's faith is mentioned at the beginning but implicitly dismissed as having anything to do with his science. *Cosmos* shows us his impoverished family saying grace at the dinner table and explains that he "took [their] fundamentalist Christian faith to heart. It would always remain a source of strength, comfort and humility for him." That's it—nothing more than a warm blanket on a cold night.

As Casey Luskin pointed out already (Chapter 13), you can find helpful essays on Faraday and Maxwell and their respective religious beliefs by Ian H. Hutchinson, professor of nuclear science and engineering at MIT. Hutchinson fills in the missing details, for instance:

> One example of the influence of [Faraday's] theological perspective on his science is Faraday's preoccupation with nature's laws. "God has been pleased to work in his material creation by laws," he remarked, and "the Creator governs his material works by definite laws resulting from the forces impressed on matter." This is part of the designer's art: "How wonderful is to me the simplicity of nature when we rightly interpret her laws." But, as [Geoffrey] Cantor points out [in his book *Michael Faraday: Sandemanian and Scientist*], "the consistency and simplicity of nature were not only conclusions that Faraday drew from his scientific work but they were also metaphysical *presuppositions* that directed his research." He sought the unifying laws relating the forces of the world, and was highly successful in respect of electricity, magnetism, and light.[168]

As for Maxwell, his faith is not mentioned in this episode at all, though he and Faraday, while coming from very different social and economic backgrounds, were equally devout. As Stephen Meyer recounts in *Signature in the Cell*, Maxwell insisted that a verse from *Psalms*, "The works of the Lord are great, sought out of all them that have pleasure therein," be inscribed in Latin over the entrance to the Cavendish Laboratory at Cambridge University.[169]

168. I. H. Hutchinson, "Michael Faraday: Scientist and Nonconformist," http://silas. psfc.mit.edu/Faraday/ and Ian Hutchinson, "James Clerk Maxwell and the Christian Proposition," http://silas.psfc.mit.edu/Maxwell/maxwell.html.

169. Stephen C. Meyer, *Signature in the Cell* (New York: HarperOne, 2009), 145.

If *Cosmos* were simply a science textbook for high school or college students, then minimizing discussion of the personal commitments that drove the greatest scientists would be understandable. The emphasis, correctly, would be on explaining the science not elaborating other philosophical views.

But from the start, Neil Tyson has wanted to tell very personal stories of the men and women he portrays as scientific heroes. That's smart. Clearly, his intention has been to excite young viewers, sparking their passion for science. And he has done a fine job as far as that goes. My own 12-year-old son is a fan of the series and complains when we have to skip an episode, noting exactly how many installments we have missed.

Given that, there is no good reason for obscuring the place of religion in scientific discovery, and Tyson knows it. Clearly, the issue was at the very front of executive producer Seth MacFarlane's mind.[170] Tyson has gone out of his way, indeed twisting the facts, to depict faith as an obstacle to science. But when acknowledging its vital role in scientific history would be most appropriate, *Cosmos* invariably falls silent.

---

170. See Chapter 3.

22.

Cosmos & Philosophy

Neil deGrasse Tyson Isn't Only Taking Hits from Us, You Know

David Klinghoffer

A S OF EPISODE 10, COSMOS FANS WERE UPSET WITH US FOR CRITICIZ-
ing their favorite program.[171] They seemed to think if only *Evolution News & Views* stopped harping on it, Neil deGrasse Tyson would have no detractors. Not true!

Massimo Pigliucci and Damon Linker were concerned about a *Nerdist* podcast that Tyson did back in March 2013 where, in a jokey way, Dr. Tyson seemed to dismiss the value of philosophy as so much navel-gazing.[172] As Pigliucci points out, he's said much the same in other forums that were less jocose. Pigliucci is at pains to emphasize what good buddies he and (the more famous) Neil Tyson are:

> It seems like my friend Neil deGrasse Tyson has done it again: he has dismissed philosophy as a useless enterprise, and actually advised bright students to stay away from it. It is not the first time Neil has done this sort of thing, and he is far from being the only scientist to do so. But in his case the offense is particularly egregious, for two reasons: first, because he is

171. Dan Arel, "Creationists grow increasingly desperate in feud with Neil deGrasse Tyson," *Salon* (14 May 2014), http://www.salon.com/2014/05/14/creationists_grow_increasingly_desperate_in_feud_with_neil_degrasse_tyson_partner/.
172. Katie Levine, "Neil DeGrasse Tyson Returns Again," *Nerdist* (7 March 2014), http://www.nerdist.com/pepisode/nerdist-podcast-neil-degrasse-tyson-returns-again/.

a highly visible science communicator; second, because I told him not to, several times.[173]

Damon Linker, meanwhile, hurls insults in defense of the great philosophers: "[Tyson] is a philistine." "He should be ashamed of himself."[174]

Frankly, I'm not all that anxious to hear Neil Tyson philosophize. A person in public life should be judged not by how he does in the tasks he doesn't set for himself, but how he does in the tasks he does set for himself.

With *Cosmos*, Tyson seeks to explain, with particular attention to the younger audience, the history of science especially as it pertains to origins, cosmic and biological. Executive producer Seth MacFarlane said from the beginning that the program has intelligent design in its crosshairs.[175]

Tyson is an amiable man. It's hard not to be charmed by him. But *Cosmos* conceals from viewers the truth about the role faith has played in the development of science. It tells a false story. And it has refused to acknowledge that the scientific question of origins is a lot more ambiguous, complicated, and interesting than the media generally admit.

Let Tyson tell the whole truth about these things, history and science, the self-assigned ambit of the program that he hosts. Let him do the job he actually undertook. Then if there's time left over, and there won't be, I'd like to know what he thinks about philosophy.

173. Massimo Pigliucci, "Neil deGrasse Tyson and the value of philosophy," *Scientia Salon* (12 May 2014), http://scientiasalon.wordpress.com/2014/05/12/neil-degrasse-tyson-and-the-value-of-philosophy/.

174. Damon Linker, "Why Neil deGrasse Tyson is a philistine," *The Week* (6 May 2014), http://theweek.com/article/index/261042/why-neil-degrasse-tyson-is-a-philistine. A philistine is: "a person who is hostile or indifferent to culture and the arts, or who has no understanding of them."

175. See Chapter 3.

EPISODE 11:
THE IMMORTALS

23.

SECULAR RELIGION

THE FIRST EPISODE OF *COSMOS* THAT
I WOULDN'T SHOW MY KIDS

David Klinghoffer

CASEY LUSKIN WILL HAVE MORE DETAILED COMMENTARY ABOUT EP-
isode 11 of *Cosmos*—see the next chapter—so I won't steal his thun-
der. I'll simply say—whoa, that's one I won't be watching with my 12-year-
old. My oldest son and I enjoyed the series together up till this point. By
chance, we missed this episode the night that it aired so I watched it online
myself.

With past installments I had to pause now and then to point out to
Ezra where host Neil deGrasse Tyson had stopped talking about science
and switched to baiting Christianity. Episode 11 would make that proce-
dure impractical and tiresome.

Almost the whole thing is a screed—executed artfully as always, but a
screed nonetheless.[176] This is no longer an exploration of the cosmos. It's
startlingly heavy-handed indoctrination in an alternative secular religion,
intended to displace any other faith-related notions in the minds of the
program's audience. That audience is clearly imagined as being young and
impressionable.

Casey and I were talking about where, with two episodes left and having dropped the initial pretense of a program on science, the Reverend Dr.

---

176. Thoughtful reader William questioned my use of the word "screed" here. OK, fair
enough. How about "sermon"?

Tyson will go from here? I found that my own Ship of the Imagination officered no assistance in guessing.

24.

PANSPERMIA

PANSPERMIA, ENVIRONMENTAL ALARMISM, SOCIALISM, GAIA, NAZI-COMPARISONS, AND MORE: COSMOS'S ENDGAME BECOMES CLEAR

Casey Luskin

WITH 11 OF 13 TOTAL EPISODES OF *COSMOS* UNDER OUR BELT, THE overall arc of the series is becoming clear. The first few episodes bashed religion and promoted materialism, while of course advocating that life developed by a process of "unguided" or "mindless" evolution. Then, for a few episodes, the anti-religious rhetoric was toned down a little, and *Cosmos* focused more on simply presenting good, uncontroversial science. But the final few episodes seem poised to ramp up the propaganda to levels not seen before.

Episode 11 pushes a naturalistic origin of life and the Copernican principle (the idea that Earth is insignificant in the cosmic scheme)—which is perhaps to be expected. But the episode gets surprisingly ideological as well, promoting panspermia, the Gaia hypothesis, and a propagandistic, *Star Trek*-like picture of the future. According to *Cosmos*, this last can only be achieved if we embrace an alarmist environmental vision. Our host, Neil deGrasse Tyson, compares skeptics of the current "consensus" on climate change to Nazis.

In the second episode of *Cosmos*, Tyson admitted, "Nobody knows how life got started." In this episode he re-tackles the topic. Tyson again says "nobody knows" how life arose, *yet all of the ideas he is willing to entertain are entirely naturalistic*. He very briefly suggests that "perhaps it [life] began in a shallow sunlit pool," or maybe "life could have started in the searing heat of a volcanic vent on the deep sea floor." And just how did this happen? Here's what Tyson tells us: "Somehow, carbon-rich molecules began using energy to make copies of themselves."

That's right: the explanation is "somehow." It reminds me of the scene from the film *Expelled*[177] where the explanation given by materialists for the chemical origin of life was "whatever it was."

All this comes just after Tyson has explained how living organisms contain a "language that all life on Earth can read" complete with a "code" that is "written in an alphabet" with "letters" where "each word is three letters long," and carries a "message" that is "copied." Lest viewers suspect any of this points to intelligent design, Tyson immediately reassures us that "everything is a masterpiece written by nature and edited by evolution." But when he asks "Where did that message come from?" he's forced to answer: "Nobody knows." For a guy who says he doesn't know how life arose, he seems strangely confident that it happened naturalistically.

But the message conveyed by this episode really isn't that "nobody knows" how life arose. In fact, Tyson spends a lot of time promoting his own pet hypothesis for the origin of life.

After some passing references to Earth-based models of the origin of life (which of course omit any mention of intelligent design as a possibility), Tyson devotes two lengthy segments of the episode to panspermia—the idea that life arrived on Earth from space—and the existence of alien life. He speculates that "life came to Earth as a hitchhiker" on meteorites, and even asserts, "There's a good bet that our microbial ancestors spent some time in space."

Panspermia: Soft, Hard, and Directed

Yes, there are a few people out there in the scientific community who agree with Tyson's viewpoint. Ideas about panspermia are periodically batted about—with varying degrees of seriousness and plausibility.

---

177. "Expelled—Origins of Life," http://www.youtube.com/watch?v=AOPkXFTd5Rs.

Some support a *soft panspermia* hypothesis, where life might be transferred between planets within our solar system. Fewer advocate *hard panspermia*, where life can cross interstellar space, spreading from star system to star system. Under this scenario, life from a single planet might ultimately seed an entire galaxy.

Tyson unequivocally promotes *hard panspermia*. In his view, life arose somewhere in the galaxy, and then was passed from solar system to solar system via supernova explosions which eject microbe-carrying meteorites through vast regions of interstellar space. Eventually, he explains, due to the orbit of star systems around the galaxy, some of these chunks of life-bearing matter land on a planet, and life gets started in a new location.

This is definitely not the mainstream scientific view. Tyson is welcome to think what he wants, but he never mentions that his opinion isn't widely shared. Indeed, there are good scientific reasons why it isn't.

The main problem with panspermia comes in explaining how life arose in the first place. Generally, advocates of the notion treat this as a problem that doesn't really need to be addressed. Why? Because the whole idea behind panspermia is that if life exists throughout the entire galaxy, then tracing its origin might be next to impossible. Thus we can defer the question indefinitely. Life might have arisen under rare, ideal conditions on some long-lost planet whose sun went supernova billions of years ago. But you could never know for sure what happened, so advocates of panspermia find it convenient to focus elsewhere. An article in *Scientific American* makes this point:

> Enthusiasts for panspermia go further, and have been known to invoke these mechanisms for galaxy-wide dispersal of life—taking one rare occurrence of life and spreading it across the stars. In some ways the motivation for proposing this kind of cosmic panspermia is a little dated. It comes from a time when we felt that the origin of life of on Earth was such a mystery, and such an unlikely event, that it was convenient to outsource it. Although this didn't actually solve the real question of life's origins, it meant that a specific origin "event" could be extremely rare among the 200 billion stars of the Milky Way yet life would still show up in other places.[178]

178. Caleb A. Scharf, "The Panspermia Paradox," *Scientific American* (15 October 2012), http://blogs.scientificamerican.com/life-unbounded/2012/10/15/the-panspermia-

This was basically Francis Crick's view. He advocated not just panspermia but *directed panspermia*—where the seeding of solar systems was directed by intelligent aliens. Just as the *Scientific American* article suggests, Crick adopted this view after he realized the great difficulties faced in explaining the origin of life on Earth. As Crick famously wrote, "An honest man, armed with all the knowledge available to us now, could only state that in some sense, the origin of life appears at the moment to be almost a miracle, so many are the conditions which would have had to have been satisfied to get it going."[179]

Problems with Panspermia

BUT EVEN IN THE QUESTIONS THAT PANSPERMIA PURPORTS TO ANswer—like how life spread from planet to planet, or star system to star system—the model has major problems. Panspermia must address (at the very least) the following three problems:

1. How could life survive ejection events, and impact events?
2. How could life survive in the harsh environment of space?
3. What vehicle could successfully transport life to new planets or star systems?

Tyson spends only a little time dealing with these issues. As I mentioned already, he never admits how skeptical many scientists are that the difficulties can be solved.

For example, regarding the first problem, Edward Anders wrote in *Nature* that "organic matter cannot survive the extremely high temperatures (>104 K) reached on impact, which atomize the projectile and break all chemical bonds."[180] Tyson doesn't mention this.

As to the second problem, Tyson proposes that life might spread not just within our solar system, but from star system to star system, traveling

---

paradox/.

179. Francis Crick, *Life Itself: Its Origin and Nature* (New York: Simon and Schuster, 1981), 88.

180. Edward Anders, "Pre-biotic organic matter from comets and asteroids," *Nature* 342 16 November 1989), 255–57, http://www.nature.com/nature/journal/v342/n6247/abs/342255a0.html.

across the entire galaxy. But Paul S. Wesson of the University of Waterloo observes that most scientists think life couldn't survive such a trip:

> The majority opinion is that while organisms may be ejected from an Earthlike planet by the collision of an asteroid or comet, the DNA and RNA is so degraded in space that the probability of seeding life in the Galaxy is low....
>
> The only feasible way to avoid this conclusion [that radiation would kill organisms in space] is to step away from the "standard" version of panspermia, and argue that organisms are transported not in dust grains but in large boulders. This is possible in principle, and can work in practice inside the solar system (see below); but is disfavored for transport over larger distances by statistical arguments.[181]

In other words, panspermia faces a catch-22. Space is vast, and the odds of life finding its way to another planet in another star system are extremely low. If life was carried on small dust particles, this might allow a great enough dispersal to increase the odds. But dust particles don't provide the necessary protection to prevent interstellar radiation from killing life.

Now life might be protected from radiation if it's hidden inside larger boulders (i.e., meteorites). But these chunks of rock are too big to be accelerated by solar wind, and (as we'll discuss below) are so rare that the chances of one of them finding its way to another planetary home outside a solar system like ours are essentially nil. Wesson writes:

> The probability of life being carried to other clusters, or across the Milky Way, is accordingly very small indeed. Thus while the transport of living organisms inside boulders may be viable within a solar system like ours, it is unlikely in a Galactic context, and is disfavored compared to the traditional version of panspermia with radiation-driven dust grains.[182]

Wesson concludes: "On a statistical basis, panspermia with living organisms must therefore be regarded as somewhat unlikely."[183] Similarly, as *Nature News* explains, panspermia is only "possible, provided that the bugs don't have to travel too far: They would probably be sterilized by cos-

---

181. Paul S. Wesson, "Panspermia, Past and Present," *Cornell University Library*, http://arxiv.org/ftp/arxiv/papers/1011/1011.0101.pdf. See page 1.

182. Ibid, 11.

183. Ibid, 12.

mic rays and UV radiation during a journey from another solar system."[184] Likewise, a *Nature* blogger explains that mainstream science is highly skeptical of this hypothesis:

> Does this mean that panspermia could have seeded life across the galaxy? Not quite. There's the enormous distance between stars, and a timescale in the tens of millions of years for a trip. The longer the period of time, the less likely life is to survive a trip to another world. *In the mind of mainstream science, this makes interstellar panspermia an unlikely occurrence.*[185]

As for the third problem—how life could be transported from one star system to another—this is probably where panspermia faces its biggest obstacles. In a famous paper in the journal *Astrobiology*, Professor H. Jay Melosh explains why it's extremely unlikely that a life-bearing meteorite ejected from a solar system like ours would ever find its way to another planet:

> The overall conclusion is that it is very unlikely that even a single meteorite originating on a terrestrial planet in our solar system has fallen onto a terrestrial planet in another stellar system, over the entire period of our solar system's existence. Although viable microorganisms may be readily exchanged between planets in our solar system through the interplanetary transfer of meteoritic material, it seems that the origin of life on Earth must be sought within the confines of the solar system, not abroad in the galaxy.[186]

He ran Monte Carlo simulations and performed calculations to determine the likelihood that an ejected meteorite would be captured by another star system, and then once in the star system would strike a terrestrial planet. He concludes:

> However, when the probability of 1 in 10,000 that the captured meteorite actually strikes a terrestrial planet is factored in, it seems unlikely that any rock ejected from a terrestrial planet in our solar system has ever reached a terrestrial planet in another solar system. This conclusion validates the quote from Carl Sagan that opened this paper. It is also in good agreement

---

184. Philip Ball, "Alien microbes could survive crash-landing," *Nature* (2 September 2004), http://www.nature.com/news/2004/040902/full/news040830-10.html.

185. Bruce Braun, "Life Traveling in Space: A Story of Panspermia," *Scitable* (11 September 2013), http://www.nature.com/scitable/blog/postcards-from-the-universe/life_traveling_in_space_a. Emphasis added.

186. H. J. Melosh, "Exchange of Meteorites (and Life?) Between Stellar Systems," *Astrobiology* 3 no. 1 (2003), 207, http://www.lpl.arizona.edu/~jmelosh/InterstellarPanspermia.pdf.

with the fact that no hyperbolic meteorites or comets have ever been observed. The spaces between the stars are immense, and the probability of exchanging material with another stellar system is correspondingly tiny.[187]

And what was that conclusion of the great Carl Sagan? It was this: "It is unlikely that a single meteorite of extrasolar origin has ever reached the surface of the Earth."[188] Melosh writes:

> When the long transit times from one star to another are added in, the prospect that life hopped from star to star by any natural agency becomes vanishingly small. The bottom line is that the origin of life on Earth must be sought within the confines of the solar system itself, not abroad in the galaxy.[189]

Though Tyson heavily promotes panspermia, he acknowledges none of these objections. Instead, he triumphantly touts an experiment where bacteria that lived on a satellite that orbited the Earth for a few years were "still alive and kicking when they were brought back to Earth." That's interesting, but it's a far cry from spending millions of years in interstellar space, unprotected from the harsh radiation outside our solar system.

Tyson simply asserts: "After thousands of years, fragments of the rocks ejected from Earth can fall as meteorites into the atmospheres of newborn planets." He says we might "imagine this process repeated from world to world, each one bringing life to others. Life would then propagate like a slow chain reaction through the entire galaxy. This could be how life came to Earth." Tyson adds, "We do not know for sure," but the consensus *thinks it does know*, and it has concluded that it didn't happen like that. Tyson is entitled to disagree with the majority of scientists, but he should be candid about the challenges to his view.

In the end, we must ask: Does Tyson's hypothesis really solve the mystery of life's origin? Of course it doesn't. It just pushes the question back into a mysterious, as-yet unexplored region of space where we can only hope that the conditions were more favorable for generating life than on Earth. In an article in *Nature*, Philip Ball explains that panspermia is rejected by "most scientists" because it's probably untestable:

187. Ibid, 214.
188. Ibid, 207.
189. Ibid, 214.

Some scientists have speculated that these molecules are not home-grown at all—that life was seeded from space, by spores carried deep-frozen through the interstellar emptiness from a living world elsewhere. This idea, called "panspermia," was first mooted in 1907 by the Swedish chemist Svante Arrhenius, and was revitalized in the 1960s by Francis Crick, the co-discoverer of the structure of DNA.

But ultimately this is not only unsatisfactory as a hypothesis, off-loading the central question to another place, but unscientific—because it's not obvious how it could ever be tested. Most scientists prefer to assume that the molecules that constituted the earliest organisms arose from simpler, small molecules formed by non-biological processes on the young Earth.[190]

Simon Conway Morris also summarizes the consensus view:

The idea that we might represent marooned colonists—perhaps from a long-dead planet engulfed in some stellar catastrophe—has a romantic appeal that taps a recurrent root in humans of displacement and longing. Not, of course, that these hypothetical colonists would be anything more than bacteria or some such equivalent. In any event, the history of life provides no evidence (although perhaps it should) of any subsequent visitation, let alone intervention, by extraterrestrials. Of course, getting even bacteria across interstellar wastes, those cubic parsecs of hard vacuum drenched in radiation, is in itself so problematic that it may be reasonable to suppose that if panspermia (that is, transport from one star system to another) occurs at all it can only be by a directed, that is, an intelligent activity. And this is what Crick and Orgel suggested...[191]

In other words, if panspermia is going to work, it will require intelligent design. Isn't that ironic?

After finishing his discussion of panspermia, Tyson ramps up the ideology big time. He proposes that there are "thousands of planets of other stars" gifted with life, and due to our radio signals sent into space, "they could already know that we're here." In what comes next, he:

+   Shows factories spewing pollutants into the atmosphere, disaster zones, oil rigs apparently exploding in the ocean, and dead oil-covered birds.

---

190. Philip Ball, "Special feature Part one: Origins of life," *Nature* (22 April 1999), http://www.nature.com/news/1998/990422/full/news990422-8.html.

191. Simon Conway Morris, *Life's Solution: Inevitable Humans in a Lonely Universe* (Cambridge: Cambridge University Press, 2005), 26. Leslie Orgel is a British chemist.

- Promotes a pseudo-socialistic view, claiming (wrongly) of our economic system that it assumes resources are "infinite," and lamenting that it is "profit driven" and thus only focused on "short-term gain."

- Alludes to the Gaia hypothesis, calling Earth a "tiny organism," and later stating, "The planet is now a self-sustaining intercommunicating organism."

- Promotes the Copernican principle,[192] envisioning a future where "we take the vision of the 'pale blue dot' to heart and learn how to share this tiny world with each other," and naively assuming that wide acceptance of the notion would unite humanity, meanwhile ignoring the fact that in truth we live on a privileged planet.[193]

- Strongly promotes a form of environmental alarmism, pointing to the "scientific consensus that we're destabilizing our climate," and claiming that those who disagree with him are "in the grip of denial" and in a "kind of paralysis."

Notice the irony. Tyson rejects the consensus against panspermia, while failing to disclose that fact. Yet he uses the "consensus" on global warming as a club to bully dissenters. After claiming that climate dissenters are "in the grip of denial," Tyson says: "Being able to adapt our behavior to challenges is as good a definition of intelligence as I know." So if you don't agree with Tyson's views about global warming and the policies that are necessary to fix it, then you're either not intelligent, or you're not using your intelligence. Instead, you're in "denial."

For myself, I'm very open to the possibility that the "consensus" on climate change is correct. But I can't condone the use of propagandistic labels and epithets like "science denial" to shut down discussion on an important topic.

What happens next in *Cosmos* is thus both sickening and immensely hypocritical. Tyson shows scenes of crowds cheering for Adolf Hitler and the Nazis. He says, "Human intelligence is imperfect, surely, and newly

---

192. David Hughes, "Review of *The Privileged Planet*," *Discovery*, http://www.discovery.org/a/3235.

193. "*The Privileged Planet*," http://www.youtube.com/watch?v=JnWyPIzTOTw.

arisen. The ease with which it can be sweet-talked, overwhelmed, or sub-verted by other hard-wired tendencies sometimes themselves disguised as the light of reason is worrisome." Again, the not-so-subtle message is that if you are a skeptic of what he calls the "scientific consensus that we're de-stabilizing our climate," then you are like a Nazi-follower, or perhaps a Holocaust denier.

Tyson is correct to warn against the use of propaganda, and the misuse of scientific authority, to mislead the public. Indeed, recall the falsehoods that this series has advanced—in the form of both untrue claims and con-spicuous omissions about the role of religion in the history of science. Neil deGrasse Tyson laments that the truth can be twisted by those driven by an agenda. He thinks he's innocent of that charge himself. He certainly is not.

EPISODE 12:
THE WORLD SET FREE

25.

A Warming Earth

When Ideology Trumps Science: Neil deGrasse Tyson and Cosmos on Global Warming

Jay W. Richards

The 12th, penultimate episode of Cosmos, "The World Set Free," stands out from most previous installments. Just when we've grown accustomed to circuitous narrative threads with obscure segues from one topic to the next, we get an episode that focuses like a laser on one subject—global warming.

Of course, since the series is called Cosmos, the producers needed to justify devoting one-thirteenth of the total running time to such an Earth-bound topic. So the story begins, not with Earth but with Venus.

"Once there was a world," host Neil deGrasse Tyson tells viewers, "not so very different from our own." We see footage of attractive seascapes and landscapes, which represent what Venus might have looked like during the first billion years of its existence. It was apparently pleasant for a while, like the big island of Hawaii. But then, bad things started happening—volcanoes, lava, smoke—which released heaps of carbon dioxide into the atmosphere, setting off a runaway greenhouse effect. "Once things began to unravel," Tyson explains, "there was no way back." This, not the proximity of Venus to the sun, is why the planet is now hot and oceanless.

From here, the script writes itself. On Venus, most of the carbon dioxide is in its atmosphere, which makes the planet's surface boiling hot. In contrast, most of the carbon on Earth has for eons been trapped in its rocks and moderated by the oceans. Our comfy abode has had just the right amount of carbon dioxide in its atmosphere so that it's not too hot, not too cold. Until, that is, we started "belching" that extra carbon into the atmosphere by burning coal and oil. Around the turn of the twentieth century, the amount of atmospheric carbon dioxide exceeded three hundred parts per million (ppm), and has continued to go up since then (it's now near 400 ppm). "The Earth," Tyson says, "has seen nothing like it in three million years."

Cue worrisome visuals and narration. Lots and lots of smoke stacks apparently spewing carbon dioxide into the air (most of what we see is probably steam but let that pass). As a result, it's warmer now than in the nineteenth century. The oceans are heating up. The ice caps are melting. Permafrost near the Arctic Circle is melting, allowing plants to grow where they haven't grown for a long time (apparently that's bad). Rising sea levels are giving rise to floods, including floods in which water flows down stairs inside buildings. There will be mass extinctions. The planet will look browner.

And why? Because we're careless and greedy. That's not a paraphrase. Here's Tyson: "Our carelessness and greed put all of this at risk."

Fortunately, there's a solution. After hearing a superficial argument for why the sun can't be responsible for the recent warming, we learn that we've had the solution at hand since the beginning of the industrial revolution, namely, solar energy. "The sun isn't the problem, but it is the solution." French inventor Augustin Mouchot showed how to use solar energy to heat steam way back in the 1870s.[194] But since the price of coal plummeted, Mouchot's funding was cut off. Later, cheap oil continued to crowd out opportunities for solar energy.

As if that weren't enough, there's also a lot of wind we can harness. "Wind farms take up very little land," especially since we can put windmills on the ocean.

---

194. "19th Century Solar Energy Engines," *SolPower People*, http://solpowerpeople.com/ solar-energy-engines-in-the-19th-century/.

Perhaps you think that such a radical transition to new forms of energy is unworkable, but Tyson assures us that it can be done.[195] We've done great things in the past, after all, even when our motivations weren't noble. For instance, Cold War competition led us to develop rocket technology to deliver deadly nuclear warheads to the other side of the globe. But that same technology allowed us to go to the moon. And from that vantage, we looked back and "realized our community."

"There are no scientific or technical obstacles to protecting our world and the precious life that it supports," Tyson insists. "It all depends on what we truly value. And if we can summon the will to act."

The closing visuals are "Jetsons green." We see the Arctic ice cap expanding, the Earth getting greener. There are fancy new solar facilities and wind farms. There are bountiful fields growing on top of skyscrapers in a futuristic city that looks like it might have those cool flying cars (though none are seen). For narration, we hear the inspiring 1961 moon speech by President John F. Kennedy: "We choose to go to the moon. We choose to go to the moon in this decade and do the other things, not because they are easy, but because they are hard."

For the viewer ignorant of the issues, it's probably inspiring stuff. For the well-informed critical thinker, however, it's frustrating. I'm tempted to provide a line-by-line response to this episode, but I'll stick to a few of the most glaring problems.

As with previous offerings, this episode provides a decent scoop of straightforward science and visuals smothered with a gravy-boat-sized helping of ideological bias and conventional wisdom. In this case, the ideology is not the tired science vs. religion trope to which we've grown accustomed in this series. It's the current fashion for catastrophic, human-induced climate change scenarios. (Well, last year's fashion. Apparently the producers didn't know that they're supposed to talk less about "global warming" and more about "climate change" or "climate disruption.")

---

195. Philip Dowd, "Capture the Sun & Power America with Solar–Is There a Business Case?," *Watts Up With That?* (31 July 2014), http://wattsupwiththat.com/2014/07/31/capture-the-sun-power-america-with-solar-is-there-a-business-case/.

And once again, Tyson and the producers have chosen the risk-free company of straw men over a rough-and-tumble encounter with real critics.

Throughout the episode Tyson responds to skeptical objections that virtually no serious person raises. For instance, he dutifully explains that we know about the past composition of Earth's atmosphere because we can analyze ice cores. But all the key players in the debate know about ice cores and paleoclimatology.

He squanders several minutes arguing that human activity is the source of recent carbon dioxide increases in the atmosphere. Again, this is a widespread point of agreement.[196]

The Greenhouse Effect

MORE SIGNIFICANTLY, TYSON BURNS UP VALUABLE AIRTIME EXPLAINING the historical and scientific details of the greenhouse effect. He even returns to the point near the end, referring to the patron saint of the *Cosmos* series, Carl Sagan, whose PhD thesis dealt with the runaway greenhouse effect on Venus, and who warned us of the dangers of a similar runaway greenhouse effect on Earth.

Virtually no one active in the climate debate disputes the basic physics of carbon dioxide and the greenhouse effect. The most obvious reason for such a long digression is to give naïve viewers the impression that climate change "deniers" don't understand basic physics and chemistry.

In reality, almost the entire debate about "climate change" among the scientifically literate centers on how significant the effect of carbon dioxide is on global climate. In particular, how much warming should be expected as a result of adding carbon dioxide to the atmosphere? By itself, we could predict about one degree of warming centigrade for every doubling of carbon dioxide in the atmosphere. Read that again. *For every doubling.* In contrast, human beings are adding some two parts per million per year. By itself, then, adding carbon dioxide to the atmosphere won't lead to catastrophe. That's the basic physics.

196. "CO2 Page," *Watts Up With That,* http://wattsupwiththat.com/reference-pages/ atmosphere-page/co2-page/.

There is even a case to be made that rising carbon dioxide combined with a concomitant modest warming is a net social benefit.[197] In fact, satellite measurements made over the past thirty years show that the Earth has measurably greened as a result of the rising level of carbon dioxide.[198]

So why all the panic? It's because most climate models assume that there are all sorts of *positive feedbacks* that magnify the effects of adding carbon dioxide to the atmosphere, at least when that happens over short, human-scale periods of time.

Tyson provides one such example, but without making the distinction clear between direct climate forcing and feedbacks: the reflective properties of water, land and ice. All things being equal, ice reflects more light back into space than either land or water does. So, if added carbon dioxide in the atmosphere warms the surface a bit and melts some ice, then the subtraction of ice will itself lead to less warming light energy being reflected back into space. And hence, more warming will result.

But just as there can be positive feedbacks, so too can there be *negative feedbacks*. These are effects that mitigate rather than magnify the direct effects of the carbon dioxide. Certain types of cloud cover seem to qualify.[199]

The real debate centers on the proper understanding, knowledge, and potential ignorance of the various feedbacks. So *Cosmos* ignores it entirely.

The climate models we hear so much about assume that most of the feedbacks are positive, and so they have predicted greater warming as more carbon dioxide finds its way into the atmosphere.

Observed Reality versus Computer Models

BUT SPECULATIVE PREDICTIONS BASED ON MODELS CAN ONLY GET YOU SO far. What we want to know is what's actually happening. The short answer

---

197. Roy W. Spencer, "The Social Benefit of Carbon: $3.5 Trillion in Agricultural Productivity" (18 October 2013), *Roy Spencer*, http://www.drroyspencer.com/2013/10/the-social-benefit-of-carbon-3-5-trillion-in-agricultural-productivity/.

198. Randall J. Donohue et al, "Impact of CO2 fertilization on maximum foliage cover across the globe's warm, arid environments," *Geophysical Research Letters* (19 June 2013), http://onlinelibrary.wiley.com/doi/10.1002/grl.50563/abstract.

199. Roy W. Spencer, "Oceanic Cloud Decrease since 1987 Explains 1/3 of Ocean Heating," *Roy Spencer* (3 October 2013), http://www.drroyspencer.com/2013/10/oceanic-cloud-decrease-since-1987-explains-13-of-ocean-heating/.

is that the models predict continued warming, but reality hasn't cooperated. There's been little to no measured warming since the late 1990s.[200] Some continue to deny this,[201] but most catastrophists concede the point and offer ad hoc reasons for why the warming isn't observed.[202] The point, however, is that the models that predict a global hothouse are increasingly out of sync with observations.[203]

Tyson claims just the opposite: "The observed warming is as much as predicted from the measured increase of carbon dioxide. It's a pretty tight case." Secondary predictions aren't cooperating either. Although viewers are told the Arctic ice caps are disappearing (and that is just what was predicted), that hasn't panned out.[204]

Add to that the evidence from paleoclimatology. We know that the Earth has at times had vastly more carbon dioxide in its atmosphere, and yet no runaway heating took place. There are lots of complicating factors of course, but it *suggests* that Earth's climate has some way to compensate.

I'm not claiming there's absolutely nothing to worry about. Honest people can disagree on the question of climate change. It involves about a hundred different issues, many of which simply don't admit to certainty. I am saying that an intellectually honest treatment of the subject would inform viewers of the full range of evidence and experts' views. It would admit that the predictions and the observations have diverged, and would allow for the possibility that there's something wrong with the models.

200. Antony Watts, "Global Temperature Update–Still no global warming for 17 years 10 months," (2 August 2014), *Watts Up With That?*, http://wattsupwiththat. com/2014/08/02/global-temperature-update-still-no-global-warming-for-17-years-10-months/.

201. Peter Gleick, "'Global Warming Has Stopped'? How to Fool People Using 'Cherry-Picked' Climate Data," *Forbes* (5 February 2012), http://www.forbes.com/sites/petergleick/2012/02/05/global-warming-has-stopped-how-to-fool-people-using-cherry-picked-climate-data/.

202. Anthony Watts, "List of excuses for 'the pause' now up to 29," *Watts Up With That?* (30 July 2014), http://wattsupwiththat.com/2014/07/30/list-of-excuses-for-the-pause-now-up-to-29/.

203. Anthony Watts, "95% of Climate Models Agree: The Observations Must be Wrong," *Watts Up With That?* (10 February 2014), http://wattsupwiththat.com/2014/02/10/95-of-climate-models-agree-the-observations-must-be-wrong/.

204. "Global Sea Ice Area," http://arctic.atmos.uiuc.edu/cryosphere/IMAGES/global.daily. ice.area.withtrend.jpg.

Solar and Wind

THE INTELLECTUAL DISHONESTY CONTINUES IN THE TREATMENT OF SO-lar and wind energy. This section is fingernails-on-a-chalk-board for any-one with basic knowledge of the physics and economics of energy. Obvi-ously if solar and wind energy were superior to oil, gas, and coal, we'd all be using it, even if, as Tyson seems to think, most of us were motivated entirely by greed.

That hasn't happened because, for most uses, solar and wind are very dilute sources of energy, and require huge energy collectors for measly output. The amount of the sun's energy reaching the planet's surface is no more relevant than the amount of energy bound up in the Earth's crust. What matters is how much of that energy we can access, and at what cost.

Those windmills anchored to the ocean floor are extremely expensive to build and maintain, and provide comparatively little energy. And then only when the wind is blowing. That doesn't mean we should ignore these technologies. We can and should seek to innovate. Unless we get some rad-ical tech breakthroughs, however, wind and solar simply cannot, and will not, replace hydrocarbons for most energy uses anytime in the near future.

This is "Intro to Energy 101" stuff. The fact that Tyson and the produc-ers of *Cosmos* blithely ignore all of it is yet one more example of how much they are willing to allow a biased ideology to trump scientific evidence. That's not the ideal way to bring science to the masses.

EPISODE 13:
UNAFRAID OF THE DARK

26.

MOB RULE

THE FINAL INSTALLMENT OF *COSMOS*
IS THICK WITH IRONY

David Klinghoffer

THE 13TH AND FINAL INSTALLMENT OF *COSMOS* IS THICK WITH IRO-
ny. The episode, "Unafraid of the Dark," focusing on mysteries of
the cosmos like dark matter, dark energy and cosmic rays, necessarily
functions as a summary of the themes that host Neil deGrasse Tyson has
sought to communicate. It's like the paragraph at the end of an article that
starts, "In conclusion."

The episode is Dr. Tyson's opportunity to really preach it to his intended audience of impressionable young people. He doesn't disappoint. Science, he tells us, is all about thinking for yourself and feeling okay about
saying you don't know something.

It is, he says, "one of things I love about science; we don't have to pretend
we have all the answers." Dark energy, he explains at one point, is "merely
a code word for our ignorance." Copping to that ignorance is good, because
"Pretending to know everything closes the door to knowing what's really
there." And again, "It's okay not to have all the answers."

Skepticism and humility are his watchwords. Tyson lists a series of com-
mandments: "Question authority," "think for yourself," "question yourself,"
"don't believe anything just because you want to," "test ideas," "follow the
evidence wherever it leads," "reserve judgment," and "remember, you can be
wrong." Who around here would argue with him?

Invoking his predecessor Carl Sagan's image of the Earth as an obscure "pale blue dot" becomes an occasion to punch the theme of our own insignificance, and the "the delusion that we have some privileged position in the universe."

Humility is a wonderful thing. So is resisting the impulse to assume you've got everything all figured out. So is insisting on your right to think for yourself. But in practice, upholders of the materialist view of reality— which has been the primary take-home message of the revived *Cosmos*— police our culture by punishing all these wonderful things.

The excellent science writer Philip Ball has an essay in *Aeon* taking a look at the scientific preference for beautiful simplicity in explanations, the way scientists may confuse an aesthetic preference for the simple and clear with the way reality actually is. He gives natural selection as an illustration:

> Might it even be that the marvelous simplicity and power of natural selection strikes some biologists as so beautiful an idea—an island of order in a field otherwise beset with caveats and contradictions—that it must be defended at any cost? Why else would attempts to expose its limitations, exceptions and compromises still ignite disputes pursued with near-religious fervor?[205]

In the context of biology, Darwinists persecute free thinkers—those too modest to think science has got it all figured out when it comes to explaining the apparent design of life—with what really is a "near-religious fervor." Likewise, if to a lesser degree, in cosmology.

Lashing those who doubt current scientific authority, on evolution or climate change, has been a major project of the folks who produced *Cosmos*. Executive producer Seth MacFarlane said as much before the first episode aired.[206]

As Casey Luskin writes in Chapter 20, for the makers of *Cosmos*, the only good controversy is a dead controversy, one that was settled by science long ago. Current debates on hot topics like intelligent design or global warming are deplorable.

---

205. Philip Ball, "Beauty ≠ Truth," *Aeon* (19 May 2014), http://aeon.co/magazine/world-views/beauty-is-truth-theres-a-false-equation/.
206. See Chapter 3.

The Library of Alexandria

THE CONCLUDING EPISODE OPENS WITH TYSON GIVING A TOUR OF A beautifully computer-generated model of the great Library of Alexandria in Egypt. Dr. Tyson strolls about, examining scrolls by great science thinkers of the ancient past.

He recounts how "mobs" burned the library with its contents—a huge distortion and simplification of the historical record. (What else is new? The Wikipedia article is instructive.[207]) Tyson then asks darkly, "What will happen next time the mob comes?"

If you've seen the rest of this series, you'll have no doubt that by the "mob" he means skeptics who in fact subscribe to the values of doubt and questioning that he lists as vital to science, "Question authority," "think for yourself," "question yourself," and the rest.

For such crimes, Darwin doubters and climate skeptics are relentlessly mocked, and worse, by adherents of Neil Tyson-style materialism. If there's a mob today—silencing debate, resisting the publication of dissenting views, setting arbitrary limits on the acquisition of scientific knowledge—it's them, not us.

---

207. "Destruction of the Library of Alexandria," Wikipedia, http://en.wikipedia.org/wiki/Destruction_of_the_Library_of_Alexandria.

27.

PRIVILEGED PLANET

COSMOS FINALE TAKES ONE LAST SHOT AT THE "DELUSION THAT WE HAVE SOME PRIVILEGED POSITION IN THE UNIVERSE"

Casey Luskin

THE FINAL EPISODE OF COSMOS MAKES A FITTING END, IN KEEPING with what we've seen already in the series. Much of it covers uncontroversial science, such as how cosmic rays were discovered, or why cosmology developed concepts like dark matter (to help explain why stars orbit so quickly at the edge of their galaxies) and dark energy (to help explain why the universe continues to expand despite all the matter it contains).

Neil deGrasse Tyson rightly acknowledges that ideas about dark matter and dark energy are really a "code word for our ignorance." A NASA website puts it this way:

> **What is Dark Energy?** More is unknown than is known. We know how much dark energy there is because we know how it affects the Universe's expansion. Other than that, it is a complete mystery....

> **What is Dark Matter?** ... We are much more certain what dark matter is not than we are what it is.[208]

Alongside such material we get the accustomed promotion of scientism and materialism, and especially the Copernican Principle—the idea that

---

208. "Dark Energy, Dark Matter," *Astrophysics*, http://science.nasa.gov/astrophysics/focus-areas/what-is-dark-energy/.

the universe was not designed, and that we in no sense have a privileged existence within it.

This final episode thus includes a lengthy segment quoting Carl Sagan from the original *Cosmos* series giving his famous pale blue dot monologue.[209] Sagan called Earth a "mote of dust suspended in a sunbeam" and "a lonely speck in the great enveloping cosmic dark." The monologue promotes the materialistic view that, "In our obscurity, in all this vastness, there is no hint that help will come from elsewhere to save us from ourselves." But perhaps the most telling Sagan quote replayed in the episode cites "the delusion that we have some privileged position in the universe are challenged by this point of pale light."

That comment from Sagan played an important role in instigating a project by Guillermo Gonzalez and Jay Richards—their book *The Privileged Planet*, which investigated whether Earth *does* have a privileged position. The new *Cosmos* entirely ignores the actual debate over whether Earth's position is "privileged" and promotes a straw-man caricature instead. It goes like this: If you think that Earth is a privileged planet, then you must think our planet is literally at the center of the universe, and you must think you have all the answers and there's no reason to engage in further investigation.

Tyson asks us to conduct a "thought experiment" where we consider all the stars with planets in the galaxy:

> Suppose on one of them there lives an intelligent species. One of the ten million life forms on that planet. And there's a subgroup of that species who believe that they have it all figured out. Their world is the center of the universe. A universe made for them. And that they know everything they need to know about it. Their knowledge is complete. How seriously would you take their claim?

He continues, stating that "our ancestors believed that the universe was made for them," and that "the architecture of our language, myths, and dreams comes from that prescientific age."

---

209. "Carl Sagan's 'The Pale Blue Dot,'" YouTube video, http://www.youtube.com/watch?v=Vtgy6uzgckc.

Yet again, *Cosmos* is rewriting history. The notion that the universe was "made for" us or that we have a special place in it isn't just some relic of the "prescientific age." On the contrary, that view was held by the founders of modern science (see Chapter 11) and it continues to be taken seriously by influential scientists today (see Chapter 13). These scientists certainly don't claim we "have it all figured out," they don't think the Earth is literally at "the center of the universe," and they certainly don't think "they know everything they need to know" or that "their knowledge is complete."

Cosmos is also flat wrong in suggesting that such ideas are scientifically unfruitful. Belief in a designed universe helped give rise to modern science. In contrast, it is the Copernican principle promoted by *Cosmos* that has led to failed scientific predictions. In Chapters 12 and 13 of *The Privileged Planet*, Gonzalez and Richards identify eight bad predictions stemming from Sagan's view:

1. Earth, while it has a number of life-permitting properties, isn't exceptionally suited for life in our solar system. Other planets in the solar system probably harbor life as well.

2. Our sun is a fairly ordinary and typical star.

3. Our solar system is typical. We should expect other solar systems to mirror our own.

4. Even if our solar system is not typical, there are lots of planetary configurations that are consistent with the presence of biological organisms. Variables like the number and types of planets and moons are mainly contingencies that have little to do with the existence of life in a planetary system.

5. Our solar system's location in the Milky Way is relatively unimportant.

6. Our galaxy is not particularly exceptional or important. Life could just as easily exist in old, small, elliptical, and irregular galaxies.

7. The universe is infinite in space and matter and eternal in time.

8. The laws of physics are not specially arranged for the existence of complex or intelligent life.

Gonzalez and Richards find that there are many parameters of our Earth, sun, moon, solar system, galaxy, and universe that make our position exquisitely well-suited to both life and scientific discovery. They conclude that when we look at the scientific evidence, weighing it in a scale with these predictions, we find that the Copernican principle

> actually may have slowed the progress of science, by leading astronomers to underestimate the importance for life of seemingly trivial details like comets, asteroids, moons, and outlying planets. Similarly, it may have discouraged astronomers from giving the concept of our solar system's habitability zone due credit.[210]

To appreciate how Sagan's viewpoint has hindered scientific discovery, consider what Sagan wrote in the book, *Cosmos*, published a few years after the original 1980 *Cosmos* series aired. Here's how he articulated the Copernican principle:

> we live on an insignificant planet of a humdrum star lost between two spiral arms in the outskirts of a galaxy which is a member of a sparse cluster of galaxies, tucked away in some forgotten corner of a universe.[211]

In reality, nothing could be further from the truth. Our Milky Way galaxy is flat and disk-shaped with spiral arms. At its center is a giant black hole that rips apart any star system that gets too close. The area around the galactic core is densely packed with stars and filled with intense radiation that would destroy Earth's atmosphere and any life. The center of the galaxy is clearly not a desirable location.

On the other hand, a position too far from the center would also be inhospitable to life because the outskirts of the galaxy lack sufficient heavy elements necessary for complex life. The optimal location for life within our galaxy is a narrow band in the middle that escapes the large zones of deadly radiation at the core, yet contains the necessary elements. This region, called the galactic habitable zone, is precisely where our solar system resides.

The very concept of the galactic habitable zone was developed in part by Guillermo Gonzalez. It supports his reasoned conviction that the cos-

---

210. Guillermo Gonzalez and Jay W. Richards, *The Privileged Planet: How Our Place in the Cosmos is Designed for Discovery* (Washington D.C.: Regnery, 2004), 256.

211. Carl Sagan, *Cosmos* (New York: Ballantine, 1985), 159.

mos was designed, and that Earth occupies a privileged position within it. That's good science.

Our distance from the center and our position between the galactic arms are also important. Were our solar system located inside the arms, extreme radiation from supernovae and "star nurseries" would again be a problem for life. Contrary to Dr. Sagan's belief that we are "lost between two spiral arms," we are placed exactly where a life-friendly solar system needs to be.

Earth's position in the galaxy is privileged in other ways, too. Our location isn't just optimal for life; it also provides an ideal position to view and learn about the universe. Spiral arms are full of dust and light that, much like city lights and clouds, would obscure astronomical observation. Between the spiral arms, our planet has a clear view of not just the galaxy but much of the universe.

The zones in the galaxy that are optimal for habitability and for astronomical observation match very closely. Despite all of Sagan's belittling remarks about our position in the galaxy, were it not for our privileged location, none of us—including Sagan—would have existed, much less be able to study the stars.

Dogmatic Answers

IN THE FINAL EPISODE, TYSON SAYS: "THAT'S ONE OF THE THINGS I LOVE about science. We don't have to pretend we have all the answers." Yet over the course of 13 episodes, *Cosmos* has repeatedly sought to give answers to the greatest metaphysical questions facing mankind.

1. The first episode quoted Carl Sagan: "The cosmos is all that is, or ever was, or ever will be."

2. The second episode said we are the result of "mindless" and "unguided" evolution.

3. The third episode portrayed humanity as "an abandoned baby on a doorstep," with no idea how we got here, and no idea "how to end our cosmic isolation." Tyson said we looked for "special meaning" in our world, but whenever we think we've found something "sa-

cred," then we "deceive ourselves and others." He told viewers that Isaac Newton's religious studies "never led anywhere," and said that Newton's appealing to God is "the closing of a door. It doesn't lead to other questions."

4. Episode 4 promoted the multiverse hypothesis with only the brief qualification that it is "speculative."

5. In Episode 5, Tyson wrongly cast the Chinese philosopher Mozi, an early scientific thinker, as being "against faith" when in fact he was a monotheist.

6. In Episode 6, Tyson said, "The most revolutionary innovation of all to come to us from this ancient world, was the idea that natural events were neither punishment nor reward from capricious gods. The workings of nature could be explained without invoking the supernatural."

7. Episode 11 promoted panspermia (without acknowledging its flaws), the Gaia hypothesis, an alarmist environmental mindset and a propagandistic, *Star Trek*-like picture of the future, comparing current skeptics of the "consensus" on climate change to Nazis.

You don't have to take my word for it. The creators of *Cosmos* have been admirably clear about their agenda.

In an interview with Bill Moyers, Tyson admitted that *Cosmos* has larger, non-scientific goals, stating that we must "think of *Cosmos* not as a documentary about science," but rather about "why science matters" and why "science is an enterprise that should be cherished as an activity of the free human mind." He referred to the show's hoped-for impact on "these states of mind that you carry with you for the rest of your life." And what are those "states of mind"? When asked by Moyers whether faith and reason are compatible, he answered, "I don't think they're reconcilable," and later stated: "God is an ever-receding pocket of scientific ignorance."[212]

Executive producer Seth MacFarlane said in an interview with *Esquire*: "There have to be people who are vocal about the advancement of knowl-

212. "Full Show: Neil deGrasse Tyson on the New *Cosmos*," *Moyers and Company* (10 January 2014), http://billmoyers.com/episode/full-show-neil-degrasse-tyson-on-the-new-cosmos/.

edge over faith."[213] Executive producer Brannon Braga is creator of numerous *Star Trek* episodes. There's nothing wrong with that—in fact I'm a big fan of his work.[214] Yet during a talk at an International Atheist Conference in 2006, Braga described his involvement in *Star Trek* as creating "atheist mythology." He stated his "conviction that religion sucks, isn't science great, and how the hell can we get the other 95% of the population to come to their senses?" He even said *Star Trek* provides a "template for a world," where "religion has been vanquished, and reason drives our hearts"—a future he says he "longs for."[215]

Cosmos appears to be part of his attempt to achieve these goals. He said in an interview that the new series aims to combat "dark forces of irrational thinking," adding that: "Religion doesn't own awe and mystery. Science does it better."[216]

This is the essential message of *Cosmos*—religion leads into "darkness," whereas only science offers truth. Such scientism is a corollary of Sagan's view that the "The cosmos is all that is," and that Earth is "an insignificant planet of a humdrum star lost between two spiral arms in the outskirts of a galaxy which is a member of a sparse cluster of galaxies, tucked away in some forgotten corner of a universe." According to *Cosmos*, only by embracing these truths can we escape the confines of ignorance that entrap us. However, every premise of this ideology is wrong:

- Earth is not lost between two spiral arms in the outskirts of the galaxy, but occupies a privileged position.

- The life-friendly fine-tuning and the finite nature of our universe suggests that our cosmos is not all that there is.

- Religious faith is not only conducive to scientific discovery, but the Judeo-Christian tradition helped give birth to modern science.

- Science is great, but it's not the only way to discover truth.

213. Stacey G. Woods, "Hungover with Seth MacFarlane," *Esquire* (18 August 2009), http://www.esquire.com/features/the-screen/seth-macfarlane-interview-0909.

214. "Brannon Braga, *Wikia*, http://en.memory-alpha.org/wiki/Brannon_Braga.

215. Brannon Braga, "Star Trek as atheist mythology," YouTube video (25 June 2006), http://www.youtube.com/watch?v=iJm6vCs6aBA.

216. Marshall Honorof, "Rebooting Cosmos," *Yahoo News* (14 January 2014), http://news.yahoo.com/rebooting-39-cosmos-39-neil-degrasse-tyson-explains-131522449.html.

Tyson is right to say that "Pretending to know everything closes the door to finding out what's really there." But throughout this series, *Cosmos* has exemplified exactly the sort of complaisance that its host condemns. There is a genuine scientific controversy going on about materialism, but *Cosmos* has not in any way sought to objectively investigate the positions in that debate.

It has given the appearance of investigation, but in fact the series has consistently whitewashed both the scientific and the historical evidence, evidence that shows materialism to be a false picture of reality. That's too bad. It's a disservice to science, and to the program's intended audience.

CONCLUDING OBSERVATIONS

28.

Lies as Useful Myths

Darwin-Defending Historians Debate Whether It's Justified for *Cosmos* "to Lie" for the Sake of Science

Casey Luskin

THERE SHOULDN'T EVEN BE A DEBATE OVER WHETHER, IN DEFEND-
ing truth, it's ever justified to lie. Yet as *Cosmos* aired, an international consortium of scholars and teachers interested in the humanities and
social sciences—including Darwin-defending historians of science—were
mulling whether it's acceptable for the series "to lie" for a good cause—in
this case, defending the authority of science. They posted their comments
at H-Net.org for all to see.

In a post titled "We need to talk about *Cosmos*," historian of science Joseph Martin,[217] who now teaches at Michigan State University, referred to
the falsehoods promoted by *Cosmos* about the history of science.[218] *Cosmos*,
as the reader of this book will know, persistently offers a false version of
history where religion never positively influences the development of science. Martin writes:

> I've been watching with interest as the history of science community,
> particularly on Twitter, has reacted with consternation to the historical
> components of Neil deGrasse Tyson's *Cosmos* reboot. To a large extent I
> agree with these criticisms. It is troubling that the forums in which the

217. "Joseph Martin," *H-Net*, https://networks.h-net.org/users/joseph-martin.
218. For examples, see Chapter 6.

public gets the most exposure to history of science also tend to be those in which it is the least responsibly represented.

But part of me also wants to play devil's advocate. First, *Cosmos* is a fantastic artifact of scientific myth making and as such provides a superb teaching tool when paired with more responsible historical presentations and perhaps some anthropological treatments of similar issues like Sharon Traweek's *Beamtimes and Lifetimes.*

Second, I don't know that we, as a community, have adequately made the case that the scholarly view of history we advance is, in fact, more useful for current cultural and political discourse than the naïve view scientists advance. One thing we often see in our research, and parallel work in philosophy of science, is that "right" is often not the same thing as "useful." I'm interested in generating some discussion in why and how, if at all, we can make the case that "useful" and "right" are and should be the same thing in this case for reasons other than internal professional ones.[219]

Let me translate. First, he acknowledges that *Cosmos* has been legitimately criticized for its inaccurate portrayal of the history of science.[220] But he wants to defend *Cosmos*, playing the "devil's advocate." Why? Because the "naïve view scientists advance"—that science is always good, and religion is always getting in the way—might be more "useful" when talking to the public, even if it isn't "right." But what does he mean by "useful"? And is he really suggesting it might be acceptable to lie in the service of defending the prestige of science? Yes he is, and that becomes clear in his next comment:

If we [grant] *Cosmos* the artistic license to lie, the question is then whether it [is] doing so in service of a greater truth and if so, what is it? And what does it mean for us if it turns out that *Cosmos* and the history community are simply going after different truths?

For the record, I myself am still very much on the fence about this issue, but if I were tasked with mounting a defense of *Cosmos* as it stands, one of the things I'd say is that the stakes of scientific authority are very high right now, especially in the United States. Perhaps the greater truth here is that we do need to promote greater public trust in science if we are going

219. Joseph Martin, "We need to talk about *Cosmos*..." *H-Net* (14 May 2014), https://networks.h-net.org/node/25318/discussions/26537/we-need-talk-about-cosmos.
220. See Chapter 12.

to tackle some of the frankly quite terrifying challenges ahead and maybe a touch of taradiddle in that direction isn't the worst thing.[221]

A "taradiddle," of course, is a lie, a statement known by its maker to be untrue and made in order to deceive. So basically he's saying it's possibly permissible for *Cosmos* to lie about the history of science, if that helps "promote greater public trust in science."

It's a sickening idea that lies may be necessary to get people to trust an institution like science. That aside, how can we know that "it"—the community that Tyson is purporting to speak for, and using falsehoods to do so—deserves our trust if it might be necessary to lie to defend it?

All this reminds me of how biology professor and science blogger Bora Zivkovic said it's all right to teach false claims to students for the purpose of "gaining trust" so they'll "accept evolution."[222] It also reminds me of an infamous comment from Darwin-defending philosopher of science Philip Quinn in Michael Ruse's book *But Is It Science?*, which discussed whether creationism is science. Quinn explains that when "good arguments fail to persuade or carry the day," then one should use "effective bad argument." According to Quinn:

> Convinced of the overall rightness of one's position, one opts to present the effective bad argument. Each time one does this, one's hands get a little bit dirtier. At first one is painfully sensitive to even small compromises that one knows to be violations of one's intellectual integrity, but gradually numbness of conscience sets in. At last, when presenting the effective bad argument has become easy and habitual—second nature, as it were—one's hands have become dirty beyond all cleansing and one suffers from a thoroughgoing corruption of mind.

He says that in order to "resist" the creationists, "it is morally permissible for us to use the bad effective argument, provided we continue to have qualms of conscience about getting our hands soiled." He assuages his conscience with the following suggestion:

---

221. Martin, "We need to talk."

222. Anika Smith, "Lying in the Name of Indoctrination," *Evolution News & Views* (27 August 2008), http://www.evolutionnews.org/2008/08/lying_in_the_name_of_indoctrin010661.html.

But I also believe we must be very careful not to allow ourselves to slide all the way down the slippery slope to intellectual corruption. Perhaps, if we divide up the labor so that no one among us has to resort to the bad effective argument too frequently, we can succeed in resisting effectively without paying too high a price in terms of moral corruption.[223]

If you read Quinn's full quote, you find that he believes that, because "creationists" are "shysters" and promote "dreadful" and "pernicious" initiatives, that means it is "morally permissible" to respond using "bad arguments" (i.e., fallacious arguments) which amount to "compromises that one knows to be violations of one's intellectual integrity" and lead to "intellectual corruption," simply because those bad arguments are "effective" in defending evolution. (In case you're wondering, the "effective" but "bad arguments" that Quinn is talking about are the dubious demarcation criteria which Michael Ruse endorsed during his testimony in the 1981 case *McLean v. Arkansas.*[224] Judge Overton then adopted and employed Ruse's demarcation criteria when ruling that creationism is not science.)

Quinn thinks one can "retreat back to the academy to wash one's moderately soiled hands." However, if the academy is recommending we should tell "taradiddles" or promote fallacious arguments simply because they're "effective," then that hardly seems like the place to go in search of moral or ethical cleansing.

---

223. Philip L. Quinn, "Creationism, Methodology, and Politics," in *But Is It Science?: The Philosophical Question in the Creation/Evolution Controversy*, Michael Ruse, ed. (Amherst: Prometheus Books, 1996), 397–399.
224. Casey Luskin, "Does Challenging Darwin Create Constitutional Jeopardy?," *Discovery Institute*, http://www.discovery.org/f/5151.

29.

Materialism for the Masses

Casey Luskin

LOOKING BACK AT THE FULL 13 EPISODES OF COSMOS, WHAT CAN WE conclude? If there were any doubt that *Cosmos* would have a materialistic message, recall the first sixty seconds of its Fox premiere on March 9, 2014.

The opening featured President Barack Obama, with the presidential seal in the background, endorsing the series and praising "the spirit of discovery that Carl Sagan captured in the original *Cosmos*."[225]

Taken alone, Obama's words are uncontroversial and politic. However, immediately following the President's statement, the show replayed Sagan's famous materialistic credo from the original 1980 *Cosmos* series, stating: "The cosmos is all that is, or ever was, or ever will be."[226]

Knowing those behind *Cosmos*, this isn't surprising. Host Neil deGrasse Tyson believes "God is an ever-receding pocket of scientific ignorance."[227] Executive producers include comedian Seth MacFarlane, who expresses his desire to be "vocal about the advancement of knowledge over faith,"[228]

225. "President Obama's Cosmos Introduction," YouTube video, http://www.youtube.com/watch?v=qcdYYISYh0I.
226. Carl Sagan, "*Cosmos* Intro," YouTube video, http://www.youtube.com/watch?v=R7n71pm0K04.
227. "Neil deGrasse Tyson Tells Bill Moyers Why Faith and Reason Are Reason Irreconcilable," *Alternet* (11 March 2014), http://www.alternet.org/news-amp-politics/neil-degrasse-tysons-new-show-cosmos-and-why-faith-and-reason-are-irreconcilable?.
228. Stacey G. Woods, "Hungover with Seth MacFarlane," *Esquire* (18 August 2009), http://www.esquire.com/features/the-screen/seth-macfarlane-interview-0909.

and *Star Trek* writer Brannon Braga, who says "religion sucks" and admits he "longs for" the day when "religion is vanquished."[229]

With Tyson himself admitting we must view "*Cosmos* not as a documentary about science," the series barely hides its ambitions to bring Sagan's materialistic views to a new generation.[230]

But have its creators pushed the agenda too far? *Cosmos* has faced sharp criticism from leading evolutionists for inventing stories about religious persecution of scientists while whitewashing religion's positive historical influence on science.

The first episode portrays the sixteenth-century "scientist" Giordano Bruno being burned at the stake by Catholic priests for teaching that the Earth orbits the sun. The problem? Bruno wasn't a scientist and he wasn't persecuted for his heliocentric views. Of course Bruno's persecution was tragic, but the church killed him for promoting the occult worship of Egyptian deities and other quirky theological beliefs.

Throughout the series, Tyson repeats this theme that religion opposes scientific advancement, whitewashing the chorus of historians who believe that religion had a positive influence on science.

As prominent historian Ronald Numbers argues, "The greatest myth in the history of science and religion holds that they have been in a state of constant conflict."[231] One scholar at the staunchly pro-evolution National Center for Science Education even blasted *Cosmos* for its "slipshod history of science" and "antireligious bias."[232]

The series's apparent goal is to inspire an atheistic vision of scientific utopia—claiming we are the result of "mindless" and "unguided" evolution, while scrubbing religion's positive contributions. Even worse, *Cosmos*

229. Brandon Braga, "*Star Trek* as Atheist Mythology," YouTube video (24-25 June 2006), http://www.youtube.com/watch?v=iJm6vCs6aBA.

230. "Neil deGrasse Tyson Tells Bill Moyers Why Faith and Reason are Irreconcilable, *Alternet* (11 March 2014), http://www.alternet.org/news-amp-politics/neil-degrasse-tysons-new-show-cosmos-and-why-faith-and-reason-are-irreconcilable.

231. Ronald Numbers, *Galileo Goes to Jail and Other Myths about Science and Religion* (Cambridge, MA: Harvard University Press, 2010), 1.

232. Peter Hess, "A Burning Obsession," *Science League of America* (14 March 2014), http://ncse.com/blog/2014/03/burning-obsession-cosmos-its-metaphysical-baggage-0015452.

brands dissenters from the "consensus" as unthinking Nazi-followers who lack "scientific literacy" and are "in denial." But is it a crime to scientifically challenge the consensus? After all, *Cosmos* heavily endorses panspermia—the fringe idea that life came to Earth from space.

The final episode pushes the idea that humanity occupies no special cosmic location, calling Earth "a lonely speck" and citing "the delusion that we have some privileged position in the universe."

Credible scientists disagree. Proponents of intelligent design have shown that Earth does occupy a privileged position that fosters both intelligent life and scientific discovery.[233] As Nobel Prize winning physicist Charles Townes explained:

> Intelligent design, as one sees it from a scientific point of view, seems to be quite real. This is a very special universe: it's remarkable that it came out just this way. If the laws of physics weren't just the way they are, we couldn't be here at all.[234]

With expensive CGI and Tyson's gifts as a science communicator, *Cosmos* offers lucid scientific explanations. But the series also shows what happens when celebrity atheists are given millions of dollars to promote their views on national television.

---

233. *The Privileged Planet*, http://www.theprivilegedplanet.com/.

234. Bonnie A. Powell, "'Explore as much as we can': Nobel Prize winner Charles Townes on evolution, intelligent design, and the meaning of life," *UC Berkeley News* (17 June 2005), http://www.berkeley.edu/news/media/releases/2005/06/17_townes.shtml.

30.

RESPONDING TO *COSMOS*:
WHY IT MATTERS

Douglas Ell

I USED TO BE AN ATHEIST. THERE WAS A TIME WHEN I WOULD HAVE cheered the religion-bashing in this remake of *Cosmos*. I would have loved the opening statement, lifted from Carl Sagan's original series, that "The cosmos is all that is, or ever was, or ever will be." I would have been pleased with the discussion of the multiverse—a nice attempt to dodge the mystery of why anything exists. (I wouldn't have minded that there is no scientific evidence to support it.) I would have bought into the recurring theme of the series that science and religion are in conflict.

The remake is pretty. It has nice graphics, and it should help get kids interested in science. Many won't notice that facts are twisted to cast religion in a negative light. Many of those who do notice won't think it important. But I can tell you from experience, it matters.

You see, I became an atheist because of similar misinformation. In high school, I read about the Miller-Urey experiment, with its suggestion that life could have formed by pure chance. Now both Miller and Urey have admitted that the mere accidental creation of amino acids (they used electric sparks through a mixture of chemicals) does not explain the origin of life, just as a pile of random letters does not explain the origin of the works of Shakespeare.

Today even Richard Dawkins admits that atheists don't have a clue how life got started. But I didn't know that then, and even today most people don't know that. Most high school textbooks still give the Miller-Urey experiment as a possible explanation for the origin of life. Anti-religious misinformation is rampant, and it matters.

It took me thirty years to climb out of the atheist hole. I learned a lot of science along the way. I learned scientists have found positive evidence of the existence of God in the form of information, in the structure of the universe, and in the DNA of every species.

There is much more. In my book *Counting to God: A Personal Journey Through Science to Belief*, I nail seven scientific challenges on the door of the Church of Atheism. The discovery that science supports belief in God has made a tremendous difference in my life. It has given me hope, comfort, and purpose. I know my life matters. And I ask you, what could matter more? Is life really only about material things and material comforts? Does the one who dies with the most toys really win? Is there no morality other than might makes right?

So I think the relationship between science and religion matters a great deal to each of us. I also think it matters a great deal to our society. We need to unite the best of science and religion. In the words of Pope John Paul II: "Science can purify religion from error and superstition; religion can purify science from idolatry and false absolutes. Each can draw the other into a wider world, a world in which both can flourish."[235]

Perhaps most of all, I think the issue of science and religion matters for our young people. About 60 percent or more of children raised with a religious faith will abandon that faith at some point. A key reason is the misplaced belief that science and faith are in conflict.

When we deny our children the true facts of science, they are vulnerable to claims that the universe is pointless and life is but a chemical accident. We watch in horror as they abuse themselves with drugs and alcohol,

235. Pope John Paul II, "Letter of His Holiness John Paul II to Reverend George V. Coyne, S.J., Director of the Vatican Observatory" (June 1, 1988), http://www.vatican. va/holy_father/john_paul_ii/letters/1988/documents/hf_jp-ii_let_19880601_padre-coyne_en.html.

and as they abuse each other. We ask "where is God?" when it happens. I think God asks why we withhold and deny evidence of design. "For since the creation of the world God's invisible attributes—his eternal power and divine nature—have been understood and observed by what he made, so that people are without excuse."[236]

Books have been written about the importance of religion, and I can't say it all here. As Alexis de Tocqueville wrote, "the soul has needs that must be satisfied." Religious people are healthier and live longer. Religious people have greater self-esteem. They have greater family and marital happiness. They are more likely to escape inner-city poverty, and they are less susceptible to depression. Religion generally inoculates people against social problems, including suicide, drug abuse, and crime.

For all of these reasons, we need to unite against false claims that religion and science are in conflict. People need to know that science—the observation, experimental investigation, and explanation of natural phenomena—overwhelmingly supports belief in God. They need to know there is an objective basis for faith. They need to know we are here on purpose, and not because of random, meaningless events. They need to know that human beings are not "chemical scum."

It is my hope that students, parents and teachers will make use of the important resources in this book to inform young people about what science really says.

---

236. Romans 1:20 (International Standard Version).

Contributors

Douglas Ell

A prominent attorney and former atheist, Douglas Ell is author of *Counting to God: A Personal Journey through Science to Belief*. Ell has lectured on the relationship between science and faith at MIT, where he double-majored in math and physics. He earned a master's degree in theoretical mathematics from the University of Maryland and a law degree from the University of Connecticut School of Law, from which he graduated *magna cum laude*. Ell drafted the first 401(k) plan in professional sports and has represented a number of nationally recognized corporations, unions, and pension plans. He has litigated nationally with great success and persuaded Congress to make important changes in employee benefits laws.

David Klinghoffer

David Klinghoffer is a Senior Fellow at Discovery Institute and the editor of *Evolution News & Views*. With Senator Joseph Lieberman, he is the co-author most recently of *The Gift of Rest: Rediscovering the Beauty of the Sabbath*. His other books include *Why the Jews Rejected Jesus*, *The Discovery of God: Abraham and the Birth of Monotheism* and the spiritual memoir *The Lord Will Gather Me In*. He is a former literary editor of *National Review* magazine and a graduate of Brown University.

Casey Luskin

Casey Luskin was trained as a scientist and an attorney, having earned his bachelor's and master's degrees in earth sciences at the University of California at San Diego and a law degree from the University of San Diego. He has conducted scientific research at the Scripps Institution for Oceanography and studied evolution extensively at both the undergraduate and graduate levels. He is Research Coordinator at Discovery Institute and co-author of the popular curriculum *Discovering Intelligent Design: A Journey into the Scientific Evidence*. He is a co-author, with Douglas Axe and Ann Gauger, of *Science and Human Origins* (Discovery Institute Press).

Jay W. Richards

JAY W. RICHARDS, PHD, IS AUTHOR OF MANY BOOKS INCLUDING THE *New York Times* bestsellers *Infiltrated* (2013) and *Indivisible* (2012). He is also the author of *Money, Greed, and God*, winner of a 2010 Templeton Enterprise Award; and co-author of *The Privileged Planet* with astronomer Guillermo Gonzalez. Richards is an Assistant Research Professor in the School of Business and Economics at The Catholic University of America and a Senior Fellow at Discovery Institute. In recent years he has been Distinguished Fellow at the Institute for Faith, Work & Economics, Contributing Editor of *The American* at the American Enterprise Institute, a Visiting Fellow at the Heritage Foundation, and Research Fellow and Director of Acton Media at the Acton Institute.

INDEX

Made in the USA
San Bernardino, CA
23 October 2014